Color Tab Index to Bird Groups

WATER BIRDS

WATERFOWL — Geese, Swan, Ducks · 4–23

SEABIRDS — Loon, Grebe, Pelicans, Cormorant, Anhinga · 24–29

HERONLIKE BIRDS — Bittern, Herons, Egrets, Ibises, Crane · 30–45

MARSH BIRDS — Rail, Gallinule, Coot · 46–49

SHOREBIRDS — Plovers, Sandpipers, plus Woodcock · 50–74

GULL-LIKE BIRDS — Gulls, Terns, Skimmer, plus Puffin · 75–89

LAND BIRDS

HAWKLIKE BIRDS — Osprey, Kite, Eagles, Hawks, Falcons · 92–107

CHICKENLIKE BIRDS — Pheasant, Grouse, Turkey, Quail · 108–13

PIGEONLIKE BIRDS — Pigeons, Doves, plus Roadrunner · 114–20

OWLS AND OTHER NOCTURNAL BIRDS — Owls, Nighthawk · 121–28

*HUMMINGBIRDS — Hummingbirds, Swifts, plus Kingfisher · 129–37

*WOODPECKERS — Woodpeckers, Sapsuckers, Flicker · 138–45

FLYCATCHERS — Phoebes, Flycatchers, Kingbirds · 146–52

SHRIKE, VIREOS — Shrike, Vireos · 153–57

*JAYS, CROWS — Jays, Magpie, Crows, Raven · 158–64

*SWALLOWS — Martin, Swallows · 165–69

CHICKADEES, WRENS — Chickadees, Titmice, Wrens · 170–85

*THRUSHES, MIMICS — Bluebirds, Thrushes, Thrashers · 186–99

*WARBLERS — Ovenbird, Warblers, Redstart · 200–15

*TANAGERS, GROSBEAKS — Tanagers, Grosbeaks, Buntings · 216–26

SPARROWS — Towhees, Sparrows, Ju

*BLACKBIRDS, ORIOL 240–51

*FINCHES — Finches, Re

*Groups with particularly colorful b   , or orange in their plumage.

OTHER BIRD BOOKS BY DONALD AND LILLIAN STOKES

The New Stokes Field Guide to Birds: Eastern Region
The New Stokes Field Guide to Birds: Western Region
The Stokes Field Guide to the Birds of North America
Stokes Field Guide to Warblers
Stokes Field Guide to Birds: Eastern Region
Stokes Field Guide to Birds: Western Region
Stokes Guide to Bird Behavior, Volumes 1–3

· · · ·

Stokes Beginner's Guide to Bird Feeding
Stokes Beginner's Guide to Birds: Eastern Region
Stokes Beginner's Guide to Birds: Western Region
Stokes Beginner's Guide to Hummingbirds
Stokes Beginner's Guide to Shorebirds

· · · ·

Stokes Bird Feeder Book
Stokes Bird Gardening Book
Stokes Birdhouse Book
Stokes Bluebird Book
Stokes Hummingbird Book
Stokes Oriole Book
Stokes Purple Martin Book

OTHER BOOKS BY DONALD AND LILLIAN STOKES

The Natural History of Wild Shrubs and Vines
Stokes Beginner's Guide to Bats (with Kim Williams and Rob Mies)
Stokes Beginner's Guide to Butterflies
Stokes Beginner's Guide to Dragonflies (with Blair Nikula and Jackie Sones)
Stokes Butterfly Book (with Ernest Williams)
Stokes Guide to Animal Tracking and Behavior
Stokes Guide to Enjoying Wildflowers
Stokes Guide to Nature in Winter
Stokes Guide to Observing Insect Lives
Stokes Wildflower Book: East of the Rockies
Stokes Wildflower Book: From the Rockies West

THE
STOKES
ESSENTIAL POCKET GUIDE TO THE
BIRDS
OF NORTH AMERICA
DONALD & LILLIAN STOKES

Photographs by Lillian Stokes and Others

Maps by Paul Lehman

Maps Digitized by Matthew Carey

LITTLE, BROWN AND COMPANY
New York | Boston | London

Little, Brown and Company
Hachette Book Group
237 Park Avenue, New York, NY 10017
littlebrown.com

First Edition: October 2014

Little, Brown and Company is a division of Hachette Book Group, Inc.
The Little, Brown name and logo are trademarks of Hachette Book Group, Inc.

The publisher is not responsible for websites (or their content) that are not owned by the publisher.

The Hachette Speakers Bureau provides a wide range of authors for speaking events. To find out more, go to hachettespeakersbureau.com or call (866) 376-6591.

ISBN 978-0-316-01051-1
Library of Congress Control Number 2014933516

10 9 8 7 6 5 4 3 2 1

IM

DESIGNED BY LAURA LINDGREN

Printed in China

Contents

Introduction   vii

How to Use This Guide   viii

In the Field and at Home   x

Ten Ways You Can Help Save the Birds   xvii

THE BIRDS

Water Birds

WATERFOWL · 4
Geese, Swan, Ducks

SEABIRDS · 24
Loon, Grebe, Pelicans, Cormorant, Anhinga

HERONLIKE BIRDS · 30
Bittern, Herons, Egrets, Ibises, Crane

MARSH BIRDS · 46
Rail, Gallinule, Coot

SHOREBIRDS · 50
Plovers, Sandpipers, plus Woodcock

GULL-LIKE BIRDS · 75
Gulls, Terns, Skimmer, plus Puffin

Land Birds

HAWKLIKE BIRDS · 92
Vultures, Osprey, Kite, Eagles, Hawks, Falcons

CHICKENLIKE BIRDS · 108
Pheasant, Grouse, Turkey, Quail

PIGEONLIKE BIRDS · 114
Pigeons, Doves, plus Roadrunner

OWLS AND OTHER NOCTURNAL BIRDS · 121
Owls, Nighthawk

HUMMINGBIRDS · 129
Hummingbirds, Swifts, plus Kingfisher

WOODPECKERS · 138
Woodpeckers, Sapsuckers, Flicker

FLYCATCHERS · 146
Phoebes, Flycatchers, Kingbirds

SHRIKE, VIREOS · 153
Shrike, Vireos

JAYS, CROWS · 158
Jays, Magpie, Crows, Raven

SWALLOWS · 165
Martin, Swallows

CHICKADEES, WRENS · 170
Chickadees, Titmice, Wrens

THRUSHES, MIMICS · 186
Bluebirds, Thrushes, Thrashers

WARBLERS · 200
Ovenbird, Warblers, Redstart

TANAGERS, GROSBEAKS · 216
Tanagers, Grosbeaks, Buntings

SPARROWS · 227
Towhees, Sparrows, Junco

BLACKBIRDS, ORIOLES · 240
Blackbirds, Grackles, Orioles

FINCHES · 252
Finches, Redpoll, Goldfinches

Photo Credits · 261
Index · 263

Introduction

Welcome to our essential pocket guide to birds. This is designed to get you started identifying the birds in your backyard, neighborhood, nature sanctuaries, and beyond. It includes the birds you are most likely to see in these areas. It has big colorful photos showing these species in their habitat and often doing a characteristic behavior. The photos also show males and females for those species in which they look different and in flight if they are often seen flying. All photos are of adults unless captioned otherwise. The text is simplified and direct, and concentrates on just the most important features of the bird's appearance, habitat, and behavior. This guide is also a handy size that you can carry around in your pocket or keep on the windowsill overlooking your bird feeders.

Once you know the birds in this guide you will be ready for our larger guides, which contain all of the birds in North America and have far more detailed identification information. They are: our national guide, *The Stokes Field Guide to the Birds of North America,* and the regional editions, *The New Stokes Field Guide to Birds: Eastern Region* and *The New Stokes Field Guide to Birds: Western Region.* It is our hope that these guides will lead you to enjoy and appreciate North America's birds more fully throughout your life.

How to Use This Guide

START WITH THE COLOR TAB INDEX TO BIRD GROUPS

This index is on the first page. It is a list of the birds in the guide divided into groups of related birds. Following a group name like "Waterfowl" is a brief list of representative species, such as "Geese, Swan, Ducks." When you see a bird in the wild, scan down the list of groups to see if one matches. If so, turn to the page indicated and look through the group for your bird.

LOOK FOR WATER BIRDS

To further help you, we divided the bird groups into Water Birds and Land Birds. The Water Birds are the first six groups. These are generally large birds on or closely associated with water, such as gulls, ducks, pelicans, and so forth. If you cannot find a bird group that matches what you have seen but your bird is in water, look over the other bird groups under Water Birds to narrow down your choices.

LOOK FOR COLORFUL OR PLAIN LAND BIRDS

Land Birds make up most of our smaller birds and a few larger ones (like hawks and grouse). Some of these birds, such as owls, chickadees, and sparrows, have rather subdued plumage colors of black, brown, gray, and white. The rest are the small colorful birds of North America, such as warblers, buntings, hummingbirds, and finches. The colorful groups have an asterisk next to their group name.

Thus, if you see a colorful land bird, look for it among the groups with asterisks; if you see a dull land bird, look for it in the groups without asterisks.

TEXT

The text for each species gives you detailed identification information for the species (the term "length" in the heading means from tip of bill to tip of tail) and describes interesting behavior, food preferences, habitat, nest materials and placement, and vocalizations.

In addition, the Breeding Information tells you:

E: the number of eggs in a brood

I: the number of days for the incubation phase

N: the number of days for the nestling phase

B: the number of broods each year

A question mark after any of these initials indicates that the information is not known.

THE QUICK AND REGULAR INDEXES

Right inside the front cover is a quick alphabetical index to most bird names and bird groups. A complete index of all common and scientific names of birds in the guide can be found at the end of the book.

MAPS

These are particularly important, as they can give you a wealth of information on when and where you can see a given bird, but only if you know how to read them. They are quite easy.

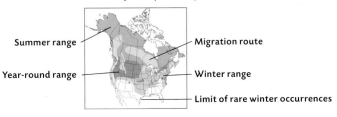

Summer range — Migration route

Year-round range — Winter range

— Limit of rare winter occurrences

Red = the summer range where the species breeds.

Blue = the winter range. If there is no blue on the map, then the species winters outside the United States and Canada.

Purple = where the species can be seen all year, both breeding and wintering. If a map is all purple, such as most of the woodpecker maps, then the species is a year-round resident and generally does not migrate.

Yellow = areas where you will see the bird only in migration during spring and fall.

Dotted Lines = While solid colors show where a bird is normally seen in the various seasons, dotted lines show where a bird may occasionally wander in a given season. The color of the dotted line shows which season, and the dotted line encloses the area where the bird may wander. You will find dotted lines only for certain birds, as not all birds wander out of their normal range.

BIRD FEEDER AND BIRDHOUSE ICONS

These are placed on the pages of birds that use bird feeders and bird-houses. We have used a generic tubular seed feeder for our symbol, but of course, different birds prefer different feeders. Hummingbirds, for example, use sugar water feeders, woodpeckers love suet feeders, and bluebirds may come to mealworm feeders. The birdhouse symbol is also generic and is not meant to represent any particular dimensions.

In the Field and at Home

BINOCULARS

Bird-watching is all about looking closely at birds, and there is no better way to do this than with a good pair of binoculars. We always carry binoculars around when outside; when inside, we keep a pair handy to see the birds out the window or at the feeder close up. There are very few things you need for bird-watching, but the two essential tools are a pair of binoculars and a field guide.

HOW TO BUY BINOCULARS

Power and Light There are two important numbers to know about when buying binoculars: these are stated as 8x42, for example. The first number indicates the power of the binoculars, and for bird-watching, this should be in the range of 7 to 8.5. Any less power and you will not see the birds as closely as you want; any more power and you may not be able to hold the binoculars steady enough for clear viewing. Some birders with steady hands do prefer 10-power binoculars. The second number indicates the diameter (in millimeters) of the lens at the far end of the binoculars. The larger this number, the more light is let in. The more light, the better you see the bird. A good range for bird-watching is 30 to 45.

Field of View Another feature to consider is the field of view, or how wide-angle your binoculars are. This can be harder to judge and compare between binoculars because it is expressed in two ways. One is by the number of degrees of a circle, such as 7 degrees; the other is by how wide the view is at 1,000 yards away, such as 345 feet. Wider angle is generally better for spotting birds, and over 7 degrees or over 300 feet at 1,000 yards will be fine for birding.

Close Focus One last point is how close the binoculars focus. Some can focus to 6 feet, and this enables you to watch butterflies, dragonflies, and warblers that come in close. Other binoculars may only focus at 12 feet, and that may be frustrating when you want to see objects closely.

Weight and Size Finally, consider weight and overall dimensions. One mistake beginners make is to buy compact binoculars, thinking that they can carry them in a pocket. But compact binoculars let in less light, have a smaller field of view, and are often hard to focus. These negatives may outweigh the positives of small size and light weight. We recommend you get full-size binoculars. Also, do some comparative shopping. Binoculars have to be just right for you, so pick them up, put the strap around your neck, and look through them. They should feel comfortable, not too heavy, and be easy to hold and focus.

HOW TO USE BINOCULARS

Almost all binoculars have eyecups that fold up and down or are extended by unscrewing slightly. If you wear glasses when looking through your binoculars, then have the eyecups folded back or screwed in. If you do not wear glasses when using binoculars, then have the eyecups folded up or screwed out. Doing this will ensure that your binoculars are the correct distance from your eye for the best view.

To spot a bird through binoculars, first fix your eyes on the bird and then carefully lift the binoculars to your eyes. If you do not see the bird, try both steps again. Once you see the bird, then move the focus knob back and forth until your image is in focus.

TIPS ON IDENTIFYING BIRDS

Identifying birds can be both fun and challenging, and can lead you to a greater appreciation of the natural world that surrounds you. Here are some tips that we hope you find helpful.

LOOKING

At the heart of identification is looking, looking closely, looking really closely. Before you even try to put a name to a new bird, look and try to describe to yourself what you see.

The more closely you look and the more specific you can be about what you see, the better birder you will be. On the last page of this guide are photos with labels showing the various parts of a bird and what each part is called. Take a moment to familiarize yourself with these photos and labels; they will help you describe birds more carefully and remember key field clues.

COLOR

Start with the bill and progress to the tail, naming every color on the bird. Look at the arrangement of the color. Is the color in spots, streaks, or patches? Then, where is the color? This is extremely important; try to be as specific as possible. For example, "I saw black on the cheeks and flanks of the bird" or "The bill was yellow at the base but black at the tip" or "There was white on the tips of the outermost tail feathers."

SHAPE

After color and its location on the bird, look at the shape of the bird, again starting at the bill and going to the tail. What is the shape of the bill? Is it long and downcurved; short, deep-based, and conical; or short, thin, and finely pointed?

Look at the length of different parts of the bird. The legs and neck—are they short or long? Is the bird deep-bellied or slim throughout? Is the tail long or short, graduated or squared-off? Overall, is the bird relatively small headed or large headed, and what is the shape of the head—triangular, round, or oval?

You can also practice noticing what we call "quantitative shape." This is where you compare the length of one part of a bird to the length of another part and come up with a comparison or ratio. For example, the bill is two times the length of the head, or the wing tips extend halfway down the tail. Thus, instead of just saying that a bird has a long tail, try to quantify its length in some way.

SOUNDS

Now use your ears. Is the bird making any sounds? If so, can you describe them in any way? Are they high-pitched or low-pitched, whistles or screeches, hoots or caws, continuous or intermittent, soft or loud, melodic or mechanical?

We have created audio CDs of all the main sounds of North American birds to help you identify the songs and calls of birds. These include: *The Stokes Field Guide to Bird Songs, Eastern Region; The Stokes Field Guide to Bird Songs, Western Region;* and *The Stokes Field Guide to Bird Songs, Eastern and Western Boxed Set.* Start by learning the songs and calls of the most common birds or those in your yard, and then expand out from there.

BEHAVIOR

Next, step back a little, look at the whole bird, and take a minute or two to watch its behavior. Describe what it is doing. If feeding, where is it looking for food? How does it get food? And what is it eating? If it is not feeding, is it relating to another bird? What actions are they doing in relation to one another? Are the birds in a flock, in pairs, or alone?

We also include the breeding behavior for each species so that you can better understand this feature of birds' lives. You can observe many things about breeding: Where is the nest built? What is it made of? How many eggs are in a brood, and how long does it takes for them to hatch?

Finally, whenever you are watching birds, always do so from a distance so as not to disturb them. Be sure never to scare them, make them take flight, or alter their behavior because of your presence. Use binoculars and telescopes to get close-up views while keeping your distance. And teach this respect for birds to others around you.

HABITAT

And finally, make a note of the habitat of the bird. Is it in a field, woods edge, backyard, shoreline, on water, in a barren lot, or other habitat?

THE CHALLENGE

As you can see, birding is a process of careful looking and discovery. You will get better at bird identification over time, but the process of discovery will continue throughout your life. There is always more to learn about any bird. In fact, a great challenge is to try to learn something new about a species every time you see it. So, instead of dismissing a common bird as "Oh, that's just a robin," take on the challenge of asking, "What can I see today in this robin that I have never seen before?"

ATTRACTING BIRDS

One of the great things about birds is that you can attract them to your property by supplying them with their basic needs: food, water, nesting habitat, and nesting material. The advantage to attracting them is that you can have the pleasure of watching them in your yard and even from your house. This is a real thrill.

BIRD FEEDERS

Bird feeders are one of the easiest ways to attract birds, and luckily, many retailers sell feeders and seed. There are several kinds of bird feeders and foods that you can offer the birds. The key is offering a variety of foods in a variety of feeders. Basic feeder types are: tubular feeders, hopper feeders, and tray/platform feeders. Tubular feeders are tubes with holes and perches near the holes. Tubular feeders with larger holes are for sunflower and mixed seed; tubular feeders with tiny holes are for thistle seed (Nyjer®), and may be called finch or thistle (Nyjer®) feeders. Tubular feeders usually attract smaller songbirds like finches, chickadees, titmice, and nuthatches. Hopper feeders may look like a house or cylinder; they dispense seeds onto a ledge. Hopper feeders attract a wide variety of birds, including larger birds like grosbeaks, cardinals, and doves, which can land on the ledge. Platform or tray feeders are open trays in which any size bird can land and access the seed.

Seed   In general it is useful to offer these kinds of birdseed: sunflower seed, mixed seed, and thistle (Nyjer®) seed.

Sunflower seed is the most popular birdseed for the majority of feeder birds. Sunflower comes as black oil sunflower, striped sunflower, and hulled sunflower, which is sunflower with the shell removed. Many birds prefer black oil sunflower over striped sunflower because black oil sunflower has a thinner shell and higher oil content. Hulled sunflower is highly desirable to birds, and will even attract birds that usually do not come to feeders, such as bluebirds, Carolina Wrens, and warblers. Another advantage of hulled sunflower is that there is no shell buildup under the feeder caused by birds that sit at the feeder and remove shells from the seed. Sunflower seed can be offered in tubular feeders, hopper feeders, and tray/platform feeders.

Mixed seed contains a variety of seeds such as sunflower, millet, milo, cracked corn, peanuts, and safflower in varying proportions. Some mixtures may contain nut pieces and dried fruits. The idea with mixed seed is that it will attract a wide variety of birds, since different birds prefer different seeds. For example, chickadees, titmice, and nuthatches prefer sunflower seed; Mourning Doves, juncos, and sparrows like millet; Blue Jays often like peanuts and cracked corn. In general, seed mixes that contain a higher percentage of black oil sunflower will be the most appealing to birds. Mixed seed can be offered in tubular, hopper, and tray/platform feeders.

Finally, there is thistle, also called Nyjer® seed, a very small, black, imported seed (not from our wildflower) that is a favorite of finches, especially goldfinches. Thistle must be offered in a special kind of tube feeder, called a finch or thistle feeder, with very small holes that keep the tiny seeds from pouring out. Note: do not put sunflower or mixed seed in a thistle feeder, since these larger seeds cannot pass through the small holes in the feeder.

Suet Feeders Suet is rendered beef fat, often sold as suet cakes, with seeds and sometimes flavors added. Since suet has a high fat content, it is very calorie rich and highly attractive to many birds, especially in winter, when birds need to consume more calories to keep warm. Suet is favored by woodpeckers, chickadees, titmice, nuthatches, and other birds. Even wrens, warblers, and bluebirds may come to suet. Suet cakes are usually offered in suet holders, which are wire cages. You can nail a suet cage to a tree trunk, hang it off a feeder pole, or use the cages that are sometimes mounted on the ends of hopper feeders. This is a very easy way to attract a great variety of birds.

Squirrel-proofing Your Feeders One of the first things people attract when feeding birds is squirrels. To keep squirrels off feeders, mount the feeder on a bird feeder pole placed at least 15 feet from anything from which a squirrel could jump. Place a squirrel baffle (a physical barrier that looks like a cone or tube) on the pole below the feeder to prevent squirrels from climbing up the pole. If it is not possible to place feeders on a pole, then buy one of the many squirrel-proof feeders on the market. These have built-in devices, such as cages or weighted perches, that prevent squirrels from accessing the seed in the feeder.

Sugar Water Feeders   Sugar water in a ratio of 1 part white table sugar dissolved in 4 parts water fairly closely reflects the composition of nectar in flowers. Hummingbirds are specialists in drinking flower nectar and so are attracted to sugar water feeders. Orioles and many other species have been recorded at sugar water feeders, and orioles may be attracted to a sugar water solution that is 1 part sugar to 5 parts water. There are two basic types of hummingbird and oriole sugar water feeders: those with a bottle filled with solution and turned upside down into a reservoir, and those that look like a saucer with a top on it. Hummingbird feeders usually have some red on them, and oriole feeders have orange.

Cleaning Feeders   For the health of the birds, clean bird feeders regularly with hot soapy water or a mild bleach solution, and rinse thoroughly. Do not let mold build up. Rake up any old seed that falls to the ground. Choose feeders that are easy to open and clean. Clean sugar water feeders and refill with fresh sugar water solution every 2 to 3 days in hot weather, since the sugar water solution can mold or degrade.

Fruit and Mealworm Feeders   Some birds will come to small trays that contain fruit, such as an orange half or grape jelly, or live or dried mealworms. Species that do not usually eat seeds, suet, or sugar water will come to these foods. We feed dried mealworms to robins, bluebirds, catbirds, phoebes, and warblers—birds not normally seen at feeders.

BIRDHOUSES

Many species of birds live in tree holes or natural cavities. Some birds, like woodpeckers, can excavate their own nest holes; others, like titmice, bluebirds, and nuthatches, need to find holes already made. Adding birdhouses to your property is a great way to attract nesting birds that use tree holes.

A wide variety of sizes of birds use nest holes, from species as large as ducks and owls to those as small as chickadees and wrens. Each size of bird needs a different-sized box with a specific entrance hole diameter.

You can learn all about which species use birdhouses and what size hole and box they need in our *Stokes Birdhouse Book*.

BIRDBATHS

All birds need water for drinking and bathing; a suitable birdbath can attract a wide variety of species. A good birdbath should have a large, shallow pool no more than 2 inches deep and with shallower water at the edges for smaller birds. It is best if the birdbath has a textured surface so that the birds can enter the water without slipping.

You can make your birdbath even more attractive by placing perches nearby. The birds can perch there and check to see if the coast is clear before bathing; they also can preen there after bathing. Adding a small bubbler or recycling fountain in the bath can attract birds; they seem to like the sound of running water.

CREATING BIRD HABITAT

With plantings and habitat management, every backyard has the potential to attract birds. The key here is to create a variety of habitats with a wide variety of plants. This will increase the range of insects, berries, and nesting habitat for your birds.

Here are a few ideas. Leave a portion of your lawn unmowed to attract crickets and grasshoppers, which birds eat. Plant a few evergreen shrubs and trees for cover at all times of year and for nesting in summer. Add berry-producing shrubs that will produce food for birds and potential nesting habitat. Add some crabapple trees with flowers for the insects they attract, fruit they offer the birds, and nesting habitat. Plant a variety of flowers. Composite flowers, like daisies and black-eyed Susans, will produce seeds for the birds. Add red tubular flowers to attract hummingbirds. And in all cases, try to utilize native plants whenever possible.

For more information see our *Stokes Bird Gardening Book, Stokes Bluebird Book, Stokes Birdhouse Book, Stokes Purple Martin Book, Stokes Oriole Book, Stokes Hummingbird Book, Stokes Beginner's Guide to Bird Feeding,* and *Stokes Bird Feeder Book.*

Ten Ways You Can Help Save the Birds

Birds need our help. In many cases, their populations are decreasing due to loss of habitat and changing climate. Anyone who has watched birds for more than a decade has seen these changes firsthand. It has sometimes become harder and harder to find birds, let alone identify them.

Here are 10 ways you can help save the birds:

1. Learn to identify more birds.

2. Learn more about bird behavior and bird needs for nesting and survival.

3. Create good habitats in your backyard by providing bird feeders, birdhouses, birdbaths, and native plantings for birds.

4. Do not use pesticides in your yard that could be toxic to birds.

5. Create a notebook of your observations of birds.

6. Share your love of birds with others, young and old.

7. Participate in bird censuses and surveys.

8. Join birding organizations and bird clubs in your area.

9. Join local, national, and international bird and conservation organizations.

10. Help your local conservation commission acquire and manage town lands so that they support more bird life, and put some of your own land under conservation easement.

The Birds

WATER BIRDS

Snow Goose

Chen caerulescens Length: 28"–31" Wingspan: 53"

These beautiful white geese with black wingtips can gather into huge flocks of hundreds to thousands during migration. When flying, they form shifting lines of white dots against the sky, and you can hear them calling from far away. They are all white except for black outer-wing feathers; the bill is pink with black edges along the opening; legs pinkish. A dark morph called the Blue Goose is mostly dark with a white head and neck. Feed on vegetation and seeds in marshes, coastal areas, and agricultural fields. Nest colonially on ground in tundra grasses. **Voice:** Well-spaced medium-pitched *aank aank.*

Breeding Info — E: 3–5 I: 23–25 N: 0 B: 1

Canada Goose
Branta canadensis Length: 27"–45"
Wingspan: 42"–60"

Our most common and widespread goose, found
feeding and breeding in a great variety of water
habitats, from rivers to lakes to city ponds. Has
distinctive black neck and head with broad white
cheeks and a grayish-brown body. Feeds on
vegetation and seeds, often grazing on lawns,
golf courses, or agricultural fields. In early summer seen in breeding pairs; otherwise usually in
large flocks. Long-distance flight is in long ever-
changing lines. Nest of grasses placed on ground
near water. **Voice:** Male a low-pitched *ahonk*;
female a high-pitched *hink*.

Breeding Info — **E:** 4–7 **I:** 28 **N:** 0 **B:** 1

Adult and young

Mute Swan
Cygnus olor Length: 60" Wingspan: 75"

A graceful swan of the Northeast and Midwest, found in water habitats from coastal to inland and freshwater to saltwater. Was introduced into North America from Europe and has now become naturalized. It is a large white swan with a long body, long pointed tail, long thin neck, and a distinctive rounded knob at the base of the bill. Often holds wings over its back in an aggressive display. Feeds on small aquatic animals and on vegetation. Nest of grasses placed on shoreline. **Voice:** Common sound is a low-pitched nasal *heeorhh*.

Breeding Info — E: 4–8 I: 35–38 N: 0 B: 1

Aggressive display

Male, domestic type

Muscovy Duck

Cairina moschata Length: 25"–31"
Wingspan: 38"–48"

Most of the Muscovy Ducks you will see are
descended from domestic birds that have naturalized; they can be found in city parks, landscap-
ing ponds, and roadside ditches. The only native
Muscovy Ducks live in southern Texas. Muscovys
are large heavy-bodied ducks that, because they
nest in tree cavities, may be seen perching in
trees. Both sexes are blackish with large white
patches on wings and sometimes on head and
neck. Male has bare facial skin and knobs at bill
base that are black to red. Female lacks knobs
and bare skin. Domestic birds vary from mostly
dark to mostly white. Feed on grain and vegetation. **Voice:** Usually silent.

Male, domestic type

Breeding Info — **E**: 8–15 **I**: 35 **N**: 0 **B**: 1

Male

Female

Wood Duck

Aix sponsa Length: 18" Wingspan: 30"

A widespread but inconspicuous duck of the shady quiet edges of wooded ponds and lakes. The gorgeous male is our most colorful duck, with a red eye and bill and a glossy green crown and crest. The female is brown with a distinctive comet-shaped white eye-ring. Wood Ducks eat aquatic plants and animals, also acorns. May be seen perched in trees because it nests in large tree holes; also nests in large birdhouses. The young leave the nest soon after hatching and simply jump to the ground without injury; the mother then leads them to the safety of water. **Voice:** Common flight call of female a high-pitched rising *oooweeek oooweeek*; male a repeated *zweeep*.

Male

Breeding Info — **E:** 10–15 **I:** 27–28 **N:** 0 **B:** 1

Male

Female

American Wigeon

Anas americana Length: 20" Wingspan: 32"

The American Wigeon is a medium-sized duck
with a short bill and pointed tail. Both sexes have
rusty flanks and a blue-gray bill with a black tip.
The male has a white crown and a broad green
swath behind its eye, also a white hip patch
and black tail; the female has a finely streaked
grayish-brown head and a pale hip and tail.
Wigeons may feed on aquatic plants and animals
or graze on grasslands. Nest of grasses can be
placed up to ½ mile from water and is concealed
by taller vegetation. **Voice:** Male gives 2- or
3-part whistling *ahweeehwoo*; female gives a low
croaking *gaahr*.

Breeding Info — E: 6–12 **I:** 23–25 **N:** 0 **B:** 1

Male

Male

Female

Mallard

Anas platyrhynchos Length: 23" Wingspan: 35"

A well-known and widespread duck from the wilds to the city center, generally preferring freshwater habitats in all seasons. The male is conspicuous and beautiful, with his iridescent green head, narrow white neck-ring, and bright yellow bill. The female is more camouflaged so as to be hidden when incubating the eggs and protecting the young. In flight, their dark blue inner wing has white borders in front and in back. Mallards eat mostly aquatic plants and animals but can also go onshore to eat acorns and grains. The nest of grasses usually near water, on ground, and under protection of foliage. **Voice:** Male gives slow drawn-out *rhaeb* and whistled *tseep*; female gives low *gwaak* and rapid *gegege*.

Males

Breeding Info — **E**: 8–10 **I**: 26–30 **N**: 0 **B**: 1

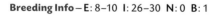

Male

Female

Mottled Duck

Anas fulvigula Length: 23" Wingspan: 30"

The Mottled Duck is most likely to be seen along
the southeast coast from South Carolina to
Texas. It is a fairly plain duck whose dark feathers
have wide buffy to reddish-brown margins; the
head and neck are contrastingly pale brown, and
they have a distinctive black mark at the base of
their lower bill that helps distinguish them from
similar female ducks. In flight, they have a dark
blue inner wing with thin white trailing edge. The
female is paler than the male. Eat aquatic vegetation as well as mollusks, crayfish, snails, and small
fish. Nest of reeds and grasses on ground near
water. **Voice:** Similar to Mallard. Male gives slow
drawn-out *rhaeb* and whistled *tseep;* female gives
low *gwaak* and rapid *gegege.*

Breeding Info — **E:** 8–10 **I:** 25–27 **N:** 0 **B:** 1

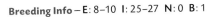

Male

Female

American Black Duck
Anas rubripes Length: 23" Wingspan: 35"

The American Black Duck winters farther north
than most of its close relatives. It lives and
breeds in fresh- and saltwater marshes. Its body
is not black but dark brownish, with a buffy
head and neck. The male has a yellow to olive
bill; the female's bill is olive with dark blotches
on the upper mandible. When flying, its white
underwings contrast strongly with its dark body.
Feeds on aquatic vegetation; in the ocean it may
eat mussels and clams. Nest of grasses placed at
shoreline. **Voice:** Like Mallard's. Male gives slow
drawn-out *rhaeb* and whistled *tseep*; female gives
low *gwaak* and rapid *gegege*.

Breeding Info — **E:** 6–12 **I:** 26–29 **N:** 0 **B:** 1

Male

Male

Female

Blue-winged Teal

Anas discors Length: 15" Wingspan: 23"

The most obvious feature of Blue-winged Teal is the large white crescent on the face of the male. The Blue-winged is named for a blue patch on the inner upper wing of male and female, seen in flight. They breed in marshes and ponds across Canada to as far south as Texas and winter in saltwater or freshwater marshes and ponds along the southern coasts. The female is like other brownish female ducks but has a whitish patch at the base of her bill. The nest is made of grasses on the ground in dense vegetation near water. **Voice:** Male gives a high whistled *wheet wheet*; female a short guttural *kut*.

Breeding Info – **E**: 6–15 **I**: 23–27 **N**: 0 **B**: 1

Male

Male

Female

Northern Shoveler

Anas clypeata Length: 19" Wingspan: 30"

The wide spatulalike bill of the Northern Shoveler is fun to see and can solve your ID very quickly. The male has an abundance of other clues, from its dark green head and black bill to its golden eye and rusty flank. The female looks like a lot of other female ducks, except for the bill. Shovelers feed at the water surface by straining out small organisms with the comblike edge of their bill; also eats duckweed and submerged plants. Nest of grasses lined with down on ground within 100 ft. of water. **Voice:** Male gives low guttural *thooktook* and *athook*; female gives short *kweg geg*.

Breeding Info – E: 6–14 **I:** 21–28 **N:** 0 **B:** 1

Male

Male

Female

Northern Pintail
Anas acuta Length: 25" Wingspan: 34"

The Northern Pintail is an elegant and handsome duck with a long slim body, long thin neck, and ski jump outline to its upper bill. The male has a particularly long pointed tail during breeding. These are ducks of open marshes and ponds in summer and coastal bays, lakes, and agricultural fields in winter. The breeding male has a brown head and throat, and a distinctive white neck that extends in a thin line up the nape. Note his white hip patch and long, thin, black tail. The female has a brown body, pale brown head and neck, and a bluish-gray bill. Eats aquatic seeds, shoots, and roots. Nest of grasses and leaves on ground usually near water. **Voice:** A liquid *tloloo* and a guttural *g'g'g'g*.

Female

Breeding Info — **E:** 6–12 **I:** 22–25 **N:** 0 **B:** 1

Male

Female

Green-winged Teal

Anas crecca Length: 14" Wingspan: 23"

One of our smallest ducks. The male combines
a reddish-brown head with a broad iridescent
green band surrounding his eye and continuing
down his nape. The female is brown like many
other females, but has a distinctive buffy patch
just under her tail. They are named for the
iridescent green patch on the trailing edge of
their inner wing, seen in flight. They breed in
freshwater ponds from Canada and Alaska to
as far south as Arizona and New Mexico, and
winter through most of the lower 48 states and
along the Pacific Coast into Alaska. Eat aquatic
vegetation and grains. Nest of grasses and leaves
on ground near water. **Voice:** Male gives a high
ringing *peep*; female a short *quack*.

Male

Breeding Info — **E**: 7–15 **I**: 21–23 **N**: 0 **B**: 1

Male

Female

Ring-necked Duck

Aythya collaris Length: 16" Wingspan: 25"

The Ring-necked is one of the diving ducks: when feeding it goes completely underwater for tubers, leaves, seeds, and mollusks and insects. Its name is a bit misleading, for its neck-ring is faint. The male is mostly dark with a whitish flank and two white rings on his bill — one at the base, the other just behind the black tip. The female's back is dark brown, her flank is warm brown, and she has a large white patch at the base of her bill in addition to the white ring near the tip. They prefer lakes and woodland ponds in summer and lakes and bays in winter. Nest of grasses and moss placed near water. **Voice:** Male a soft short whistle; female a short gruff *grek*.

Male and female

Breeding Info — **E:** 6–14 **I:** 25–29 **N:** 0 **B:** 1

Male

Female

Greater Scaup
Aythya marila Length: 18" Wingspan: 28"

This is a look-alike of the Lesser Scaup, with a greater tendency to winter in saltwater bays. Helpful clues in distinguishing the two are that the head of the Greater is generally longer than it is high, while the head of the Lesser is generally higher than it is long. The dark head of the male Greater Scaup, in good light, can appear greenish, while the head of the male Lesser Scaup, in good light, can appear purplish. The head of the Greater Scaup rounds off at the back of the crown, while that of the Lesser has a slight corner at the back of the crown. Nest of grasses lined with down placed on the ground within 200 ft. of water. **Voice:** Male a soft whistle; female a gruff *grek*.

Male

Breeding Info—E: 8–14 I: 21–28 N: 0 B: 1

Male

Female

Lesser Scaup
Aythya affinis Length: 17" Wingspan: 25"

This is another of the diving ducks, those that feed by going completely underwater. The Lesser Scaup eats tubers, leaves, seeds, mollusks and insects. The male is blackish in front and at the rear, with a gray back and pale gray flanks (flanks white in male Greater Scaup). The female has a dark brown head and neck, and grayish-brown back and flanks; she also has large white ovals at the base of either side of her bill. Greater Scaup are very similar in habits and appearance, except they tend to winter more in salt water, while the Lesser Scaup tends to winter more on fresh water and some saltwater bays. Nest of grasses lined with down placed on the ground within 200 ft. of water. **Voice:** Male a soft whistle; female a gruff *grek*.

Female and male

Breeding Info — **E:** 8–14 **I:** 21–28 **N:** 0 **B:** 1

Male

Female

Bufflehead

Bucephala albeola Length: 14" Wingspan: 21"

One of our smallest ducks, the Bufflehead is an active diver and often seen in winter in small groups on lakes or coastal waters. When actively feeding, they seem to spend as much time under water as above it, repeatedly bobbing to the surface after long dives. The male has white flanks, dark back, and a large white patch on the head behind the eye; the female is brownish-gray with a white streak across her cheek. Breed primarily in Canada and Alaska and northwestern states, and nest in tree holes or nest boxes. **Voice:** A breathy *gruuhf gruuhf*.

Breeding Info – **E:** 8–10 **I:** 28–33 **N:** 0 **B:** 1

Male

Male

Female

Common Merganser
Mergus merganser Length: 25" Wingspan: 34"

This is the largest of the three North American mergansers. It breeds and winters on open freshwater. The male has a distinctive white body, dark greenish head with no crest, and red bill; the female has a reddish-brown head, white chin, red bill, and ragged crest. Like all mergansers, they dive and catch small fish underwater. Surprisingly, they also nest in large tree holes, rock crevices, and large nest boxes along wooded lakes and rivers. **Voice:** Male a snoring *ahwuuuhn*; female a short grating *grek*.

Breeding Info — **E**: 8–11 **I**: 28–35 **N**: 0 **B**: 1

Male

Male

Female

Red-breasted Merganser
Mergus serrator Length: 23" Wingspan: 30"

This medium-sized merganser breeds across much of Canada and Alaska and winters mostly in saltwater bays along the East, West, and Gulf coasts. It is similar to the Common Merganser but the male has a ragged crest, thinner bill, and darker body; the female has a thinner bill and her reddish-brown head blends into her lighter breast. As with other mergansers, their bills have serrated edges to help in catching small fish. Nest is on ground or in a burrow or rock crevice; lined with down. **Voice:** Male a drawn-out *jadik dadeee*; female a short *uk uk*.

Breeding Info — **E**: 5–11 **I**: 29–35 **N**: 0 **B**: 1

1st-winter male

Male

Female

Hooded Merganser

Lophodytes cucullatus Length: 18" Wingspan: 24"

Our smallest merganser, most commonly seen in winter in fresh- and saltwater bays. The male is striking, with a large crest that can be fully raised to show a white center framed in black, or lowered and fairly hidden; he also has two black vertical stripes on his shoulder and reddish-brown flanks; the female is inconspicuous, with a grayish body and reddish-brown crest. They eat small fish and frogs, and breed along wooded ponds and sluggish rivers, nesting in tree cavities or large nest boxes near water. **Voice:** Male a short *grek*; female a low grating *aahwaaah*,

Breeding Info—**E**: 6–18 **I**: 32–41 **N**: 0 **B**: 1

Male

Adult summer

Adult winter

Common Loon

Gavia immer Length: 26"–33" Wingspan: 46"

The Common Loon is the quintessential symbol
of wild northern lakes, and its haunting calls
can be heard echoing out over the water day or
night. Loons are expert fishermen, snorkeling
to see the fish and then diving after their prey.
Sometimes they seem to spend more time
underwater than afloat. The adults in summer
have a checkered black-and-white back and a
black head with white barred neck-ring. In winter they are basically grayish-brown above and
whitish below. Their nest is at the water's edge
and made of reeds, grasses, and mud. **Voice:** A
haunting rising and falling wail like *tuuweeeaarr*,
a tremulous *wawawawa*, and a repeated *weeahweeep weeahweeep*.

Chick

Breeding Info — **E:** 2 **I:** 29 **N:** 0 **B:** 1

Summer

Pied-billed Grebe

Podilymbus podiceps Length: 12½" Wingspan: 16"

A small, compact, and widespread grebe that breeds in lakes and ponds over most of North America. "Pied" refers to its bill in summer, which has a black median band across it that disappears in winter plumage. In summer the bird is brownish overall, with a black chin and crown that is lost in winter, making the bird appear mostly brownish. This is a cute grebe, with a short body and tail, and a white fluffy rump. It spends a lot of time diving after aquatic insects and fish. Nest is a collection of grasses and reeds placed at water's edge, sometimes even floating. **Voice:** Call sounds a little like a giggle — *heh heh heh heh*.

Winter

Breeding Info — **E**: 4–7 **I**: 23 **N**: 0 **B**: 1

Adult spring

Adults winter

American White Pelican

Pelecanus erythrorhynchos Length: 62"
Wingspan: 108"

This is one of North America's largest birds,
found locally in large flocks within its range.
Breeds on large inland lakes mainly in the western United States and western Canada, migrates
through the West and Midwest, and winters in
the southern states and southern coastal areas.
It is all white with black flight feathers and an
orange bill. From winter to mid-breeding the
birds grow a rounded keel on their upper bill
that is then shed. They feed by dipping their bills
into the water and sieving out the small fish; this
is sometimes done by groups that seem to herd
the fish. Nest on ground with a rim of stones and
vegetation. **Voice:** Generally silent.

Adult

Breeding Info – E: 1–3 **I:** 29–36 **N:** 17 **B:** 1

Adult breeding

Brown Pelican

Pelecanus occidentalis Length: 50" Wingspan: 79"

This is our coastal pelican. It lives year-round along the coasts of southern California, the Gulf of Mexico, and the south Atlantic states. It feeds by taking short dives into the ocean to fill its pouch, then strains out the fish while emptying its pouch of water. It has a silvery body and wings, black belly, and a white head and neck; its nape turns dark brown during breeding. One- to two-year-old birds are brownish overall with whitish bellies; in their third year they look like the adult. Usually nests on coastal islands, either in trees in a stick nest or on the ground in a simple scrape. **Voice:** Generally silent.

Breeding Info—E: 2–4 **I**: 28–30 **N**: 71–88 **B**: 1

Immature

Adult

1st year

Double-crested Cormorant

Phalacrocorax auritus Length: 33" Wingspan: 52"

A distinctive waterbird of coasts, inland lakes,
and rivers. It feeds on fish by swimming after
them underwater; afterwards, comes out and
holds wings out to dry. It is named for the crests
of longer feathers that grow on either side of the
head during breeding. Adults are all shiny black
with orange facial skin and throat pouch; young
in their first two years are brownish with buffy
neck and breast. Flocks fly in long linear forma-
tions both low over the water and high up during
migration. Nest is a platform of sticks placed
near water on cliffs or on the ground. **Voice:**
Generally silent.

Breeding Info – **E**: 2–7 **I**: 24–29 **N**: 35–42 **B**: 1

Adult breeding

Breeding male

Adult female

Anhinga

Anhinga anhinga Length: 35" Wingspan: 45"

Locally called the "piano bird" and the "snake bird," the former because of the pattern of black and white feathers, and the latter because it can swim with its body submerged and just its long neck and head above water, resembling a snake. Anhingas feed on fish, which they pursue underwater and then stab with their sharp-pointed bills. They then toss their prey into the air and catch it to swallow. Males are all black with long white feathers on their upper wing; females are similar but have a tan neck and upper breast. Just before breeding, both sexes grow whitish plumes on their heads and their facial skin turns turquoise. Nest of twigs in trees. **Voice:** A snoring call.

Immature

Breeding Info — E: 1–5 **I:** 25–29 **N:** 38–42 **B:** 1

American Bittern

Botaurus lentiginosus Length: 25" Wingspan: 42"

The American Bittern is the "hidden heron," for it tends to stay within cover of marshy vegetation, posing with neck and head vertical to blend with the lines of the reeds and cattails, perfectly camouflaged with its streaked brown head and neck. It is a big, heavy-bodied heron with relatively short legs and neck. It feeds on fish, reptiles, amphibians, insects, and small mammals. Nest of cattails and grasses placed in dense marsh vegetation or tall grass fields. **Voice:** Its resonant call during breeding is a good way to discover its presence. It is a deep and repeated *gunk kerlunk, gunk kerlunk;* also gives a throaty *werk.*

Breeding Info—E: 2–7 **I:** 28–29 **N:** 14 **B:** 1

Great Blue Heron

Ardea herodias Length: 46" Wingspan: 72"

Our largest and most widespread heron, usually seen feeding alone in shallow edges of rivers, lakes, and coastal waters. Dark gray overall except for white forehead bordered by black head plumes; massive yellow bill. Slow moving and graceful, both on land and in the air. Eats a wide variety of foods, including small mammals, fish, amphibians, and reptiles. Typically breeds colonially in dead trees of beaver ponds or other swampy areas; nest is a platform of branches and twigs. **Voice:** Call is a deep, raspy *kraaar*.

Breeding Info—E: 3–7 **I**: 28 **N**: 55–60 **B**: 1

Great Egret
Ardea alba Length: 39" Wingspan: 51"

A large and graceful white heron with a bright
orange deep-based bill and long black legs. Feeds
in water through patient waiting and watching
for fish; feeds on land by slowly stalking brushy
edges as it looks for reptiles and amphibians.
During breeding its orange facial skin turns
bright green and its upper bill becomes dusky.
Its call is a short low rattle often given in skir-
mishes with other Great Egrets as they vie for
feeding areas. Nest is a platform of sticks placed
in tall shrubs or trees. **Voice:** A low, short, rattly
kraaak.

Breeding Info — **E:** 1–6 **I:** 23–26 **N:** 42–49 **B:** 1

Snowy Egret
Egretta thula Length: 24" Wingspan: 41"

The Snowy Egret is our most common, wide-spread, and beautiful white heron. It is easily recognized by its black legs with bright yellow feet or "galoshes." Black legs often have a yellow stripe up the back. Eats aquatic animals when in water and insects when on land. In the water it may shake its foot to stir up animals and then dart out its bill to catch them. The Snowy Egret is all white with a thin black bill and yellow facial skin. In breeding its facial skin and feet turn bright red, and it grows long plumes on its back, breast, and head. It nests colonially in trees and makes a platform nest of sticks. **Voice:** A low harsh *shraak*.

Breeding Info – **E**: 3–5 **I**: 20–29 **N**: 30 **B**: 1

Juvenile

Adult

Little Blue Heron

Egretta caerulea Length: 24" Wingspan: 40"

Here is a heron that is bluish as an adult but white as a juvenile. In both plumages the bill is distinctively bluish-gray at the base and dark at the tip. In its first spring and summer, the juvenile starts to change into adult plumage and can look strangely mottled. Just before breeding, the adults' bill base and facial skin turn cobalt blue. Little Blue Herons typically feed by holding their necks out at a 45-degree angle and staring for oncoming fish. It breeds colonially throughout the South, and its nest is a platform of sticks placed in tall shrubs or trees. **Voice:** A rather thin grating *shreeer*.

Breeding Info—E: 2–5 **I:** 20–24 **N:** 42–49 **B:** 1

Adult

Tricolored Heron

Egretta tricolor Length: 26" Wingspan: 36"

There is really no difference between birds called herons and those called egrets. They are just different common names for birds in the genus *Egretta*, which includes the Tricolored Heron. Wings, back, and neck are mostly slate blue. It is our only dark heron with a white belly. During breeding its facial skin and bill turn bright cobalt blue. It eats a wide variety of foods, but mostly fish, with its exceptionally long thin bill. Nests colonially in trees and shrubs; nest is a platform of twigs. **Voice:** A drawn-out harsh *creeer*.

Breeding Info — **E**: 3–7 **I**: 21–25 **N**: 35 **B**: 1

White morph

Reddish Egret
Egretta rufescens Length: 30" Wingspan: 46"

The Reddish Egret is well known for dancing about as it hunts for fish, changing directions quickly, opening its wings, and swaying its head. Although many herons engage in similar behavior to stir up prey, they do not do it as much or as vigorously. The Reddish Egret is large and has long shaggy feathers on its neck. Its bill is pink at the base and black at the tip. There is another form of this species called a "white morph." It is a wholly white version, with the same pink bill with a dark tip, and is quite beautiful. Nest a platform of sticks placed in trees. **Voice:** A grating *raaah*.

Breeding Info — E: 3–7 I: 25–26 N: 45 B: 1

Cattle Egret

Bubulcus ibis Length: 20" Wingspan: 36"

The Cattle Egret is named for its habit of following cattle and picking up insects and other invertebrates stirred up by the animals' feet. It feeds in a wide variety of open habitats, from lawns to roadsides. It immigrated to the United States in the 1950s, having originated in Africa. It is now widespread in the southern states and locally farther north. It is usually all white with a short yellowish-orange bill and black legs. In breeding it grows feathers with a buffy wash on its crown, breast, and back, and its legs and bill turn dark red. Nests colonially in trees and shrubs; nest is a platform of twigs with a lining of greenery. **Voice:** Generally silent.

Breeding Info — **E**: 2–6 **I**: 21–24 **N**: 30 **B**: 1

Adult

Black-crowned Night-Heron
Nycticorax nycticorax Length: 25" Wingspan: 44"

By day a fairly secretive and sedentary heron that becomes active at dusk and night, when it flies out to feed. As it flies from roosting locations in shubbery and trees to feeding areas in shallow water, you are most likely to hear its explosive *wok* call. The adult has a black crown and back, silvery-gray wings, and a white body; its bill is blackish. The juvenile bird is streaked brown and has a pale yellowish base to its lower mandible; this distinguishes it from the similar juvenile Yellow-crowned Night-Heron, which has an all-dark bill. Nests colonially, from the ground to high in trees. **Voice:** An explosive *wok*.

Breeding Info — E: 3–5 **I:** 24–26 **N:** 42–49 **B:** 1

Juvenile

Adult

Yellow-crowned Night-Heron

Nyctanassa violacea Length: 24" Wingspan: 42"

Of the two Night-Herons, the Yellow-crowned is the more likely to be seen by day, especially on cloudy or rainy days when the light is low. Its call as it flies out at dusk to feeding ponds is a downslurred *skeeow*. The adult Yellow-crowned has a black head with a yellow crown and white stripe on the cheek; its body is gray and its back and wings are streaked black and gray. The juvenile is streaked brown and white, and has an all-dark bill. Nests colonially in shrubs or trees; nest is a platform of twigs. **Voice:** A downslurred *skeeow*.

Breeding Info — E: 3–5 **I**: 21–25 **N**: 25 **B**: 1

Immature

Green Heron

Butorides virescens Length: 19" Wingspan: 26"

Compared to most of our other herons, the Green Heron is secretive and a loner, nesting in single pairs in the dense vegetation along the edges of marshes, lakes, and rivers. It is also generally more widespread, nesting throughout the East and Midwest and along the West Coast. When feeding, it slowly stalks its prey with great patience, then suddenly darts out its bill to catch a fish. Its crown and back are iridescent bluish-green, and its neck and breast are reddish-brown. Nest is a platform of twigs placed 5–45 ft. high in shrubs and trees. **Voice:** A sharp *keeow* and a repeated *keh keh keh*.

Breeding Info — **E**: 3–6 **I**: 21–25 **N**: 34–35 **B**: 1

Winter

Glossy Ibis
Plegadis falcinellus Length: 23" Wingspan: 36"

All of our ibises look like herons, but with a long
downcurved bill. They use their bills to probe
for crabs, worms, and insects in lawns or fields
or in the mud underwater. The adult in summer
is dark reddish-brown on head, neck, and body,
and glossy iridescent bronze-green on its wings
and tail. Its bill and legs are gray to pinkish. In
winter the head and neck are brownish, finely
streaked with white. Nests colonially in trees or
shrubs; nest is a platform of twigs. **Voice:** A nasal
wahn and *emp emp*.

Breeding Info — **E**: 3–4 **I**: 21 **N**: 28–30 **B**: 1

Summer

Adult

Adult

White Ibis

Eudocimus albus Length: 25" Wingspan: 38"

The adult White Ibis forms a striking picture in flight, with its bright red feet and bill, white body and wings, and blackish-green iridescent wing tips. Like all ibises, it flies with neck extended; herons tend to fly with their necks folded back. In breeding the legs, bill, and facial skin turn a brilliant deep red, and they develop a small gular pouch. Juvenile White Ibises are brown on back and wings in their first winter, then become mottled brown and white in their first spring as they molt into adult plumage. They nest colonially in trees and shrubs; nest is a platform of twigs with some grasses. **Voice:** A nasal honking *huuhn*.

Breeding Info — **E**: 3–5 **I**: 21–23 **N**: 28–35 **B**: 1

1st spring

Adult

Roseate Spoonbill

Platelea ajaja Length: 32" Wingspan: 50"

Roseate Spoonbills are well known for both their color and shape. They live along coastal areas of Florida and the Gulf states, mostly near mangroves and coastal islands. Their pink wings, tails, and bellies get deeper pink over the first three years of their lives, and their heads get increasingly bald and green over the same period. The bill is flat and spoon-shaped at the tip. When feeding, they wave their bills back and forth through shallow water, feeling for shrimp, fish, and other water animals. They breed colonially in mangroves or dense shubbery; the nest is a platform of sticks, sometimes with grasses. **Voice:** A grunting *uhn uhn*.

1st year

Breeding Info—E: 1–5 **I**: 22–24 **N**: 35–42 **B**: 1

Wood Stork

Mycteria americana Length: 40" Wingspan: 61"

A very large bird that can be found in southern
states. It is common locally in swamps, coastal
shallows, and roadside ditches. When on land it
appears mostly white, with a bare, gray, warty
neck and head and blackish legs with pink feet.
In flight it reveals striking all-black flight feathers
and tail. Often flies in groups, soaring high on
thermals. Feeds in shallow water of wetlands
with its head down, feeling for aquatic animals with its bill. They nest in large colonies in
cypresses or mangroves; nest is a platform of
sticks. They are an endangered species due to
loss of nesting and feeding habitats. **Voice:** Generally silent.

Breeding Info — **E:** 3–4 **I:** 28–32 **N:** 55–60 **B:** 1

Sandhill Crane

Grus canadensis Length: 41"–46" Wingspan: 73"–77"

Most people see or hear Sandhill Cranes during migration, when the birds travel between their northern breeding grounds to the southern states where they may winter sometimes in groups of thousands. You can recognize them in flight by the way their long legs and necks project out (larger herons fold their necks in flight) or by their rolling call (see Voice). They are all pale gray with rosy bare skin from the midcrown to the base of the bill. In summer they may have rusty stains on their feathers, a result of feeding in iron-rich soil and then preening. Nest a large mound of plant material placed on water or land. **Voice:** A mellow, rolling, low-pitched *kukukuroo*.

Breeding Info — **E**: 1–3 **I**: 28–32 **N**: 0 **B**: 1

Adult summer

Sora

Porzana carolina Length: 9" Wingspan: 14"

The Sora belongs to the rails, a group of secretive birds that spend most of their time on the ground in dense marsh vegetation. The Sora is the most likely to be seen, for it seems to be willing to venture out into open water and to be active during daylight. It looks a little like a small brown chicken with greenish-yellow legs and very long toes. It has a conical yellowish bill and a short triangular tail that is often flicked as the bird moves about. Soras eat mainly seeds but also go after insects and small fish. Nest is a small firm cup of woven vegetation sometimes lightly covered; it is placed above water level in marshy areas. **Voice:** A descending series of toots and a rising *kewweeet kewweeet*.

Adult winter

Breeding Info — **E**: 6–18 **I**: 18–20 **N**: 0 **B**: 1

Common Gallinule

Gallinula galeata Length: 14" Wingspan: 21"

The Common Gallinule lives in freshwater ponds and marshes. It winters mostly along the coast but also summers inland. It has a blackish body, brown wings, and a reddish bill with a yellow tip. Its bill extends up its forehead in what is called a frontal shield. Its long legs and toes enable it to walk over marsh plants as it looks for vegetation and small water animals to eat. The Common Gallinule looks similar to and may share the same waters with the American Coot. Nest a mound of marsh vegetation placed in protection of tall cattails. **Voice:** A rapid churring followed by a series of toots that seem to "run out of steam," like *churrrdootdoot doot doot dahr dahr.*

Breeding Info—**E**: 4–17 **I**: 19–22 **N**: 0 **B**: 1

Adult

Purple Gallinule
Porphyrio martinica Length: 13" Wingspan: 22"

The Purple Gallinule is one of the most spectacularly colored birds in North America. Untrue to
its name, it is not just purple, but also turquoise,
iridescent green, bright yellow, deep red, and
white. It lives at the edges of freshwater lakes
and marshes in all of the Gulf states. It is fun to
watch it walk over lily pads with its very long
yellow legs and toes delicately placed to support
it. In its 1st winter it is duller, lacking the purple
and red of the adult. It eats water vegetation and
small water animals. Nest is a floating platform
of vegetation tied to standing reeds and cattails.
Voice: A low-pitched series of muffled chickenlike clucks.

1st winter

Breeding Info — E: 5–10 I: 22–25 N: 0 B: 1

American Coot
Fulica americana Length: 15" Wingspan: 24"

The blackish body and whitish bill of the American Coot make it easy to identify. It swims in the water like a duck, but its bill is conical, not flattened, and its toes are lobed rather than webbed. Coots live in marshy areas in summer and also along the coast in winter. Pairs are aggressive around nest sites, but in other seasons coots gather by the hundreds into tight flocks in open water. Their nests are made of wetland plants and float in the shallows of marshy vegetation. **Voice:** Several sounds including a tooting *kitoot kitoot* and a grating *grrrt grrrt*.

Breeding Info—**E**: 8–12 **I**: 21–25 **N**: 0 **B**: 1

Transition to summer

Winter

Black-bellied Plover

Pluvialis squatarola Length: 11½" Wingspan: 29"

The Black-bellied Plover winters all along
both East and West coasts, and is common on
beaches and mudflats, making short runs as it
looks for food. Only during breeding, which
occurs in northernmost North America, does it
have a black belly; in winter its belly is white and
its back a mottled brown. The Black-bellied Plover is our largest plover, and about robin-sized.
In winter it eats mollusks and marine worms; in
summer, on the tundra, it eats mostly insects.
Nest is a shallow scrape on the ground. **Voice:** A
lovely, plaintive, whistled *peeaahhweeet*.

Breeding Info — **E**: 4 **I**: 26–27 **N**: 30–35 **B**: 1

Summer

Winter

Semipalmated Plover

Charadrius semipalmatus Length: 7"
Wingspan: 19"

A common small plover of mudflats and beaches,
seen in migration throughout the United States
and southern Canada and found in winter along
East, West, and Gulf coasts. Usually seen feeding
in small loose-knit flocks, either alone or with
other plovers and sandpipers. Dark brown back
and single dark breastband on white belly distinguish it from most other plovers. The similar Killdeer has 2 breastbands. Nest a slight depression
in the ground in a sandy or gravelly area. Often
best located and identified by its distinctive call.
Voice: A two-part call like *chuweet* given from
the ground or overhead.

Breeding Info—E: 4 **I**: 23–25 **N**: 0 **B**: 1

Winter

Snowy Plover
Charadrius alexandrinus Length: 6½"
Wingspan: 17"

The Snowy is our smallest plover and has a distinctively thin black bill and dark legs; otherwise, its markings are similar to those of our other small plovers. Its upperparts are dark brown to pale grayish-brown; its underparts are white. In summer it has a black (in the male) or brown (in the female) collar and forehead bar. Winter plumage is similar but collar and forehead are pale brown. Snowy Plovers eat marine worms, small crustaceans, and insects. They are primarily coastal but also nest in interior alkaline flats. Nest a shallow depression in sandy areas. **Voice:** A rising *chuweeep* and a short *jr'r'rt*.

Summer male

Breeding Info – E: 2–3 I: 25–30 N: 0 B: 1

Winter

Piping Plover
Charadrius melodus Length: 7¼" Wingspan: 19"

The Piping is a small plover with a short stubby
bill that is black with variable orange at base, and
dark orange legs. This is a very pale plover with
black collar and forehead in summer; it is similar
in winter but black areas are brown or faint. It
is an endangered species in the Great Lakes and
threatened in the northern Great Plains and
Mid- to North Atlantic coasts because its nesting habitat of coastal and shoreline areas is easily
disturbed by human recreation. Its main food is
marine worms, small crustaceans, and insects.
Nest a small depression in sand sometimes lined
with shells or pebbles. **Voice:** Named for its piping call of *peetoo*; also gives a repeated *pit pit* and
a plaintive *turwee*.

Summer male

Breeding Info — **E**: 3–4 **I**: 26–28 **N**: 0 **B**: 1

Killdeer

Charadrius vociferus Length: 10½" Wingspan: 24"

The Killdeer is the most common inland shorebird, living in open barren or short-grass fields
and mudflats. When these areas include lawns,
sports playing fields, gravel parking lots, and
even gravel rooftops, the Killdeer may come into
contact with people. If you approach its young
or nest, the adult will do the "broken wing act"
in which it tries to draw you away with loud
calls while holding its wing out to the side as if
injured. In these cases, just move away from the
bird so as not to disturb it. Its nest is just a scrape
in the ground lined with pebbles. The Killdeer
eats earthworms, spiders, and small animals.
Voice: Similar to its name, *kideer kideer, kidideer.*

Breeding Info—E: 3–4 **I:** 24–28 **N:** 0 **B:** 1

American Oystercatcher

Haematopus palliatus Length: 18½"
Wingspan: 32"

True to its name, the American Oystercatcher
really does eat oysters (along with other mollusks), prying them open with its long vertically
flattened bill. These handsome birds, with their
striking black-and-white plumage, seem to be
dressed in a tuxedo; the red bill, yellow eye, and
pink legs add a dash of color and make them
unmistakable. They live on coastal beaches and
mudflats and in salt marshes, where they scrape
out a little depression and line it with pebbles
for a nest. **Voice:** Loud penetrating whistles like
peeet peeet and a rolling trill.

Breeding Info — **E**: 2–4 **I**: 24–29 **N**: 0 **B**: 1

Male

Black-necked Stilt

Himantopus mexicanus Length: 14" Wingspan: 29"

Whether standing or in flight, the long bubble-gum-pink legs of the Black-necked Stilt are its most outstanding feature. The long legs enable the bird to wade out into deep water, where it uses its needle-thin bill to pick insects off the water surface. Stilts are mostly coastal in eastern and Gulf states but can be found on inland lakes and marshes of the Midwest and West. The bird is easily recognized by its blackish upperparts and white underparts; the female has a subtly brownish back, the male a black back. Nest is a scrape or mound of vegetation placed near water. **Voice:** A short barklike *kwit kwit*.

Breeding Info—E: 3–4 **I:** 25 **N:** 0 **B:** 1

Female

Winter male

Summer male

American Avocet

Recurvirostra americana Length: 18"
Wingspan: 31"

American Avocets are often seen in large flocks,
which magnify the beauty of a single bird. Their
long legs and delicate upturned bills give them
an elegant look, and their summer plumage of
black, white, and orange, along with their bluish legs, make them one of our most colorful
shorebirds. In winter they appear similar, but
with a white head and neck. The sexes are subtly
different in shape, the male with a longer, less
upcurved bill. Avocets usually breed in loose
colonies. The nest is a scrape or mound of vegetation placed near water. **Voice:** High-pitched,
harsh *chleeep*.

Breeding Info—E: 4 **I**: 23–25 **N**: 0 **B**: 1

Summer male

Winter

Winter

Spotted Sandpiper
Actitis macularius Length: 7½" Wingspan: 15"

The Spotted Sandpiper breeds across most of North America except for a few southeastern states. It prefers water edges, be they coastal, streamside, or lakeside. As it walks about it repeatedly bobs its whole rear as if doing the rumba. Its flight is also distinctive — very rapid fluttering in shallow arcs. The spots on its breast and belly only occur in summer. This species turns the tables, with the female sometimes mating with several males, and the males incubating the eggs and raising the young. They eat insects, crustaceans, and small fish. Nest is a shallow scrape with bits of plant material usually placed near water. **Voice:** Loud whistled *peetweet* or rapid series of *peet* calls.

Summer

Breeding Info — **E**: 4 **I**: 21 **N**: 0 **B**: 1

Juvenile, fall

Solitary Sandpiper
Tringa solitaria Length: 8½" Wingspan: 22"

The Solitary Sandpiper is an elegant midsized sand-piper that repeatedly bobs the front of its body as it walks about. "Solitary" is a good name, because this species is often found alone or in small groups, unlike many other shorebirds that are usually seen in large flocks. In summer, its back and wings are dark brown with fine white spots, its belly is white, and its neck and breast are finely streaked. It has a wide white eye-ring and yellowish-green legs. In winter, it has less distinct streaking on its breast. While breeding in the wet coniferous forests of southern Canada, it uses the deserted nest of another species, such as a grackle or robin, as a base and adds a little material. Thus it is one of the few sandpipers that nest in trees. **Voice:** High-pitched emphatic *peetweet* or *peetweetweet*.

Adult, summer

Breeding Info—E: 4 **I:** 23–24 **N:** 0 **B:** 1

Winter

Lesser Yellowlegs
Tringa flavipes Length: 10½" Wingspan: 24"

The Lesser Yellowlegs is smaller than the Greater and has a proportionately shorter bill in relation to the length of the head (see Greater Yellowlegs). In summer Lesser Yellowlegs have dark upperparts heavily spotted with white and white underparts barred on flanks but not belly. In winter they have less spotting above and less barring below. Yellowlegs are unusual among our shorebirds in that they nest in small boggy clearings in northern woods. Their nest is a shallow depression in the ground lined with fine plant material.
Voice: Differs from Greater Yellowlegs in giving only 2 loud whistles at a time, like *teer teer*.

Breeding Info — E: 4 I: 22–23 N: 0 B: 1

Winter: Greater, left; Lesser, right

Winter

Greater Yellowlegs

Tringa melanoleuca Length: 14" Wingspan: 28"

Yellowlegs are thin-bodied, long-legged shorebirds that winter along the coasts and breed
in northern forests. They are named for their
bright, school-bus-yellow legs. In summer
Greater Yellowlegs are gray with white spotting
above and white with blackish barring beneath.
In winter they have less spotting and barring.
Greater and Lesser Yellowlegs look similar
but have different proportions: the bill of the
Greater is 1½ times the head length, while the
bill of the Lesser is about the length of the head.
They feed by wading into the water and using
their fine-pointed bill to snatch insects off the
surface. Nest of leaves and grass on ground.
Voice: 3–4 loud whistles like *teer teer teer teer.*

Summer Greater, left;
winter Lesser, right

Breeding Info—E: 4 **I**: 23 **N**: 0 **B**: 1

Winter

Willet

Tringa semipalmata Length: 12"–14"
Wingspan: 26"

The Willet is a large shorebird of mostly coastal marshes and beaches, but in the West may breed in grassy wetlands or dry meadows. In winter it is found along coastal beaches and looks very plain gray above, whitish beneath, and with gray legs. But when it takes flight, bold patterns of black and white are revealed on its wings, making it easier to identify. In summer, Willets are covered with moderate to heavy brown barring. There are two fairly distinct subspecies, the Eastern Willet, which leaves the United States in winter, and the Western Willet, which winters on all coasts. Nest is a shallow depression with some vegetation. **Voice:** A loud *pawil willet, pawil willet.*

Winter

Breeding Info — **E:** 4 **I:** 22–29 **N:** 0 **B:** 1

Whimbrel

Numenius phaeopus Length: 16" Wingspan: 32"

Whimbrels are most often seen wintering or migrating along both coasts. This is a relatively large shorebird with two dark brown crown stripes and a downcurved bill that is about 2 times the length of its head. It probes into the ground to get worms or small crustaceans or picks off insects and berries in summer. Whimbrels breed mainly in Alaska and along the southern rim of Hudson Bay. Their nest is a depression in the ground lined with bits of plant material. Our only other similar shorebird is the Long-billed Curlew, but its bill is closer to 3 times the length of its head. **Voice:** Midpitched rapid series of whistles, *teeteeteetee.*

Breeding Info — **E**: 4 **I**: 22–24 **N**: 0 **B**: 1

Marbled Godwit

Limosa fedoa Length: 18" Wingspan: 30"

You are most likely to find a Marbled Godwit
in winter along coastal beaches or mudflats,
for in summer it gives up its "shore" bird status
to breed in the northern Great Plains and the
southern tip of Hudson Bay. All of our godwits
are relatively large shorebirds with distinctive,
very long, slightly upturned bills. The Marbled
Godwit is a fairly uniform cinnamon color and its
bill is bright pink at the base in winter. To feed, it
probes into the mud of tidal flats, eating worms,
mollusks, and crustaceans; in summer it eats
mostly large insects, like crickets and grasshoppers. Marbled Godwits breed in loose colonies.
Nest is a scrape in the ground. **Voice:** A repeated
harsh *kweeet*.

Breeding Info—E: 4 **I:** 21–23 **N:** 0 **B:** 1

Winter

Ruddy Turnstone

Arenaria interpres Length: 9" Wingspan: 21"

Turnstones are true to their name, for they
spend much of their time on the wrack line
of the beach flipping over stones and shells in
search of the insects and animals they eat. In
winter, they look mostly brown above and white
below with bright orange legs, but when they
take flight an exciting pattern of bold white
stripes is revealed on their wings and back. In
summer they are more colorful overall, rufous
on the wings, black and white on the head, and
with two brown loops across the chest. They
nest in the tundra in a small scrape lined with
mosses and seaweed. **Voice:** Typically a short
chert.

Summer

Breeding Info – E: 3–4 **I:** 22–24 **N:** 0 **B:** 1

Winter

Black Turnstone
Arenaria melanocephala Length: 9" Wingspan: 21"

The Black Turnstone lives on the West Coast, where it winters along the rocky coasts and breeds on the northern shores of Alaska. It specializes in eating barnacles, limpets, and insects, and places its cuplike nest of grasses in among other grasses. It is similar to the Ruddy Turnstone in having bold black-and-white stripes on its upperparts in flight, but its body is less patterned and mostly sooty brown in winter and blackish brown in summer. **Voice:** A slow *pit too wit* or a rapid mellow chattering.

Breeding Info—E: 4 I: 21 N: 0 B: 1

Summer

Summer

Winter

Sanderling

Calidris alba Length: 7½" Wingspan: 17"

The Sanderling is most often encountered on winter beaches, where it is the ultimate wave-chaser, seeming to play "catch me if you can" with each new wave. It runs out as the wave recedes, feeds frantically, and then dashes back up the beach with the next wave right at its heels. Like many small shorebirds, Sanderlings in winter are grayish-brown above and white below. In flight they have a bold white stripe through the length of their blackish wing. In summer the head and back become reddish brown. They breed in the northernmost reaches of North America, and their nests are just a shallow scrape. **Voice:** An irregularly repeated short *pit*.

Winter

Breeding Info—E: 4 **I:** 24–31 **N:** 0 **B:** 1

Winter

Least Sandpiper
Calidris minutilla Length: 6" Wingspan: 13"

The Least is the smallest of all our sandpipers. In winter it is similar to our other small sandpipers, except that the Least Sandpiper has a saving grace — its legs are yellowish. All the other small sandpipers have black legs. So if you see a very small sandpiper and the legs are yellow, you have a Least. When feeding on mudflats, Leasts tend to stay on the higher, dryer areas, rarely wading out into water like the other sandpipers. The Least tends to have a fairly well defined brown bib in summer and winter. It breeds across northern Canada and Alaska. Nest is a small depression lined with fine plant materials. **Voice:** A high-pitched, slightly ascending *kreeet*.

Breeding Info — **E**: 4 **I**: 19–23 **N**: 0 **B**: 1

Winter

Juvenile

Semipalmated Sandpiper
Calidris pusilla Length: 6¼" Wingspan: 14"

The Semipalmated Sandpiper breeds all across northernmost Canada and northern Alaska, then migrates about 2,000 miles to South America for the winter. On one of its routes south, thousands of birds take off from eastern Canada and New England and head out over the sea until they reach South America. The Semipalmated Sandpiper does not winter in the United States or Canada. Its journey north in spring is more widespread across the continent. Its summer plumage is similar to that of the Western Sandpiper but is less reddish-brown on the back and wings, and its bill is more tubular and straight, and less fine-tipped. Juveniles, often seen in fall, have buffy wash on breast and finely scaled look to upperparts. Nest is a scrape on the ground in tundra. **Voice:** A short harsh *chirt*.

Juvenile

Breeding Info—E: 4 **I:** 18–22 **N:** 0 **B:** 1

Winter

Winter

Western Sandpiper
Calidris mauri Length: 6½" Wingspan: 14"

The Western Sandpiper is most likely to be seen on coasts in winter and inland during migration. Its breeding grounds are in the tundra of northern Alaska. In summer, when compared to the similar Semipalmated Sandpiper, it has a more reddish-brown back and longer more fine-pointed bill. In winter it is most likely confused with the Least Sandpiper but has black (rather than yellow) legs and a whiter throat and breast (rather than buffy or brown). Eats insects and marine worms. Breeds in tundra; nest a scrape in the ground, usually under a shrub or grass clump. **Voice:** A high-pitched *cheet* or *cheeveveet*.

Breeding Info — E: 4 I: 20–22 N: 0 B: 1

Transition to summer

Winter

Dunlin

Calidris alpina Length: 8½" Wingspan: 17"

The Dunlin is a very common bird on beaches
in winter. It has a distinctive shape — it is fairly
deep-chested with a relatively small head, and
its bill is about 1½ times its head length, tapered
to a fine point and drooping at the tip. We often
say "D for Dunlin and D for drooping" to help us
remember this. Dunlins rapidly probe the mud
as they search for food. They look very different
in summer, with reddish-brown backs and black
bellies. They breed in the tundra, and their nests
are just a scrape in the ground. **Voice:** Fairly
drawn-out *jjeeet*.

Breeding Info — **E**: 4 **I**: 19–22 **N**: 0 **B**: 1

Winter

Winter

Short-billed Dowitcher

Limnodromus griseus Length: 11" Wingspan: 19"

The bill of the Short-billed Dowitcher is still long,
about 1½ to 2 times longer than its head. It often
uses its bill to probe into the mud in sewing-
machine fashion as it hunts for marine worms
and crustaceans. It prefers tidal mudflats and
saltwater more than the Long-billed Dowitcher
does, and is pretty much restricted to the coast
in winter. The two dowitchers are hard to distinguish by plumage and shape, but the Short-billed
does have a generally shorter bill and its call is
very different — a rapid mellow *tutututu* instead
of the high-pitched *teeek* of the Long-billed. The
nest is a simple depression in moss or grass and
is lined with fine plant materials. **Voice:** A rapid
mellow *tutututu* or *tututu*.

Summer

Breeding Info—E: 4 **I:** 21 **N:** 0 **B:** 1

Winter

Long-billed Dowitcher
Limnodromus scolopaceus Length: 11½"
Wingspan: 19"

The easiest way to distinguish between the two
dowitchers is by calls. The call of the Long-billed
is a sharp high-pitched *teeek*, while that of the
Short-billed is a mellow *tutututu*. Dowitchers
feed by rapidly probing mud with their bills,
feeling for marine worms and mollusks; at other
times they may pick off insects. The Long-billed,
during migration and winter, tends to prefer
freshwater lakes and impoundments; the Short-
billed in winter prefers saltwater mudflats. Nest
is a scrape on the ground in wet meadows near
water. **Voice:** Call is a high-pitched *teeek*.

Breeding Info — E: 4 **I**: 20 **N**: 0 **B**: 1

Winter

American Woodcock

Scolopax minor Length: 11" Wingspan: 18"

The American Woodcock is an odd-looking bird, with a deep belly, short deep neck, relatively large head, and long straight bill about 2 times the length of its head. It prefers wet meadows and thickets, where it probes the mud for earthworms. It is mostly active during feeding at dawn and dusk, and when the male does spectacular high courtship flights accompanied by twittering as it rises, chirping as it circles at the apex of its flight, and then a buzzy *peeent* when it lands on the ground. Nest a scrape in the ground lined with a few twigs and grasses, usually in the woods about 100 to 200 yards from the male's display ground. **Voice:** A buzzy *peeent*.

Breeding Info— E: 4 I: 20–21 N: 0 B: 1

SHOREBIRDS

Adult summer

1st winter

Bonaparte's Gull

Chroicocephalus philadelphia Length: 13½"
Wingspan: 33"

The Bonaparte's is a small gull with elegant patterning and proportions. Its short thin bill and rounded head give it a dovelike appearance, and its open wing reveals white outer feathers with a black trailing edge. In winter its head is white with a dark spot behind the eye; in summer it has a black head. Winters along the coasts and in southern states; nests in Canada and Alaska. Feeds on fish, dipping to the water's surface in flight, or eats aerial insects or earthworms when inland. 1st-year bird like adult winter but with mix of gray and brown on back. Unusual among our gulls, it builds its stick-and-grass nest in a conifer 5–20 ft. high. **Voice:** Alarm call a buzzy *geh geh geh*. Takes 2 years to become adult.

Adult winter

Breeding Info — E: 3 I: 24 N: 0 B: 1

Adult winter

Adult summer

Laughing Gull

Leucophaeus atricilla Length: 16" Wingspan: 40"

The Laughing Gull is the most common resident gull of the southern coast from the Carolinas to Texas; it also breeds into New England, where it is less common. It is named for its laughlike warning call, which sounds like *gagagaga*. In summer the adult has a black head and dark blackish-red bill and legs; its bill is large and appears slightly drooping. In winter its legs are blacker and its head more white with some dusky shading on top and back. 1st-winter bird has mix of gray and brown on back. Breeds colonially on beaches. Nest of grasses and sedges on ground.
Voice: High-pitched *keeeeahh keeeeahh kya kya kya*. Takes 3 years to become adult.

1st winter

Breeding Info—E: 3–4 **I:** 19–22 **N:** 0 **B:** 1

Adult summer

Adult summer

Franklin's Gull

Leucophaeus pipixcan Length: 14½" Wingspan: 36"

This is a Laughing Gull look-alike with very different behavior. Franklin's Gulls breed on the northern Great Plains, where they nest colonially in prairie marshes, feeding on aerial insects and earthworms. After breeding, most leave the United States and winter along the west coast of South America. It is a smaller version of the Laughing Gull with relatively shorter bill and legs, a less attenuated body, and, in flight, a distinctive white band separating the gray inner wing from the black tip. 1st-winter bird has mix of gray and brown on back. Nest a floating mass of vegetation anchored to cattails. **Voice:** Alarm call is a staccato *kekekeke*. Takes 3 years to become adult.

1st winter

Breeding Info—**E**: 2–3 **I**: 18–20 **N**: 0 **B**: 1

Adult winter

Adult summer

California Gull
Larus californicus Length: 21" Wingspan: 54"

The California Gull winters on the West Coast, preferring to breed and sometimes winter on lakes and rivers much farther inland. Adult has a slightly darker gray back than the paler gray Ring-billed and Herring Gulls, is midway in size between these two, and has black and red spots at the tip of the bill (Herring Gull has just red; Ring-billed has a black ring). Adult has distinctive dark eye and long evenly thick bill. In winter the California Gull may frequent dumps, parking lots, and piers, where it eats anything vaguely edible. 1st-year bird is brown with pink bill that has sharply defined dark tip. Nest a shallow depression in the ground lined with weeds and grasses. **Voice:** Alarm call is a descending *keeeow*. Takes 4 years to become adult.

1st winter

Breeding Info — E: 3 **I:** 23–27 **N:** 0 **B:** 1

Adult winter

Adult summer

Ring-billed Gull

Larus delawarensis Length: 18½" Wingspan: 48"

The Ring-billed might be called the fast-food gull, for this is the species that frequents the dumpsters and parking lots of these restaurants. Like adult California and Herring Gulls, it has a gray back and wings, and a white head and body. The adult can be recognized by the dark band near the tip of its yellow bill. This is a good species to know, since it is so widespread. 1st-year bird is whitish with brown wings and dark-tipped pink bill. Nest of grasses and pebbles on the ground. **Voice:** Alarm call a loud *kakakaka*. Takes 4 years to become adult.

Breeding Info — **E**: 3 **I**: 21 **N**: 0 **B**: 1

1st winter

Adult winter

Adult summer

Herring Gull

Larus argentatus Length: 22"–27" Wingspan: 58"

One of our largest and most common gulls with a white head and gray back. Note its pinkish legs, pale eye, and yellow bill with red spot. The Herring Gull breeds mostly in Canada, the Great Lakes, and New England along ocean shores or the edges of lakes and rivers. It frequents dumps but eats a wide variety of food, including scavenging dead fish off beaches. Its winter plumage is similar to summer but with dark streaks on the head and sometimes black on the bill tip. 1st-year bird brownish with white head, bill pinkish fading to darker tip. Nest a scrape in ground lined with grass or seaweed. **Voice:** Alarm call is a deep *keh keh keh*. Takes 4 years to become adult.

1st winter

Breeding Info – **E:** 3 **I:** 26 **N:** 0 **B:** 1

3rd winter

Adult summer

Western Gull

Larus occidentalis Length: 25" Wingspan: 58"

The Western Gull is a year-round resident of the West Coast, and it is the only common gull in the West with a dark gray back. In flight look for the adult's dark gray wings with characteristic wide white trailing edge. Its bill is relatively short and thick with a bulbous tip. 1st-year bird dark brown with all-black bill. This is the main gull that nests on the West Coast from Washington to southern California. The Western Gull is creative with nest sites, using cliffs, islands, ground, buildings, or even moored boats. **Voice:** Calls include deep throaty *kwooh kwooh kwooh*. Takes 4 years to become adult.

Breeding Info — **E**: 3 **I**: 25–29 **N**: 0 **B**: 1

1st winter

Adult fall

Adult summer

Great Black-backed Gull

Larus marinus Length: 25"–31" Wingspan: 65"

The Great Black-backed Gull is the largest common gull in the East and the only common eastern gull with a black back and wings. Always stands head and shoulders over its neighboring gulls. It is similar to the rarer Lesser Black-backed Gull but has pink rather than yellow legs. Eats wide variety of foods, even other seabirds. 1st-year birds tend to have very white heads and checkered brown or black-and-white wings and back. Nest is built on the ground and made with sticks, grasses, and seaweed. **Voice:** A very low-pitched *kwooh kwooh kwooh*. Takes 4 years to become adult.

Breeding Info — **E**: 2–3 **I**: 27–28 **N**: 0 **B**: 1

1st winter

Winter

Summer

Forster's Tern

Sterna forsteri Length: 13" Wingspan: 31"

The Forster's Tern looks similar to the Common
Tern, but there are several ways to distinguish
it. The Forster's regularly winters in the United
States; the Common mostly winters south of the
United States. Also, the Forster's is the only tern
in winter plumage to have a little black party
mask from eye to ear. In summer it can be told
from the Common Tern by its large bill and a tail
that extends beyond the wing tips on a perched
bird; the Common Tern has a shorter thinner
bill and shorter tail that does not extend past its
wing tips. During breeding the Forster's prefers
marshes; nest a mat of vegetation. **Voice:** Short
harsh *djar djar* and longer *jeeer jeer*.

Winter

Breeding Info – **E**: 3–4 **I**: 22–23 **N**: 0 **B**: 1

Summer

Summer

Common Tern
Sterna hirundo Length: 12" Wingspan: 30"

The Common Tern is most similar to the
Forster's Tern, except for several important
differences. The Common Tern generally leaves
Canada and the lower 48 states to winter from
Mexico to South America; the Forster's Tern is a
common winter resident on most of East, West,
and Gulf coasts. In summer note the subtly
grayish underparts of the Common versus the
pure white underparts of the Forster's. During
breeding, Common Terns prefer sandy to rocky
shorelines and coastal islands, while Forster's
Terns prefer marshes. **Voice:** Extended harsh
keeeh, shorter harsh *djah*, and repeated *kip kip kip*.

Breeding Info — **E**: 3 **I**: 21–27 **N**: 0 **B**: 1

Winter

Winter

Summer

Royal Tern
Thalasseus maximus Length: 18" Wingspan: 41"

The Royal, along with the Caspian, are our two largest terns. The Royal is distinguished by its shaggy black crest, which looks like it is having a bad hair day. The larger Caspian Tern has no crest and has a red bill (the Royal's is orange). The Royal Tern is strictly coastal, living year-round along the coasts of the Southeast and southern California. It eats almost entirely small fish captured through aerial dives into the water and breeds in large colonies on sandy islands. Nest a shallow scrape in the ground lined with grasses. **Voice:** Low-pitched harsh *djaar*.

Breeding Info – **E**: 1–2 **I**: 20–22 **N**: 0 **B**: 1

Winter

Winter

Summer

Sandwich Tern

Thalasseus sandvicensis Length: 15" Wingspan: 34"

Although this species is named after the town of Sandwich in Kent, England, we like to think of the bright yellow tip of this bird's black bill being the mustard on a sandwich. The Sandwich Tern is a medium-sized, sleek tern with a shaggy crest, black cap in summer, and white forehead with black nape in winter. It eats mostly fish caught from aerial dives, and it often feeds farther out to sea than many other terns. It breeds coastally, often on low sandy islands. The nest is a shallow scrape on the ground. **Voice:** Short harsh *djit djit* or longer *djaarit*.

Breeding Info—**E:** 1–2 **I:** 21 **N:** 0 **B:** 1

Winter

Winter

Winter

Caspian Tern
Hydroprogne caspia Length: 21" Wingspan: 50"

The Caspian is our largest tern; it has a heavyset body and thick-based, abruptly tapered bill. Its bright red bill and lack of shaggy crest easily distinguish it from the similar Royal Tern. It breeds coastally and inland on sandy beaches along lakes and large rivers. Its summer and winter plumages are very similar, except that its black cap is speckled with white in winter and looks like it got a buzz cut. In flight its wings are pale except for a distinctive dark underside on the outer third of its wing. Nest is a scrape in the ground lined with grasses and seaweed. **Voice:** Drawn-out raspy *schaaar*.

Breeding Info— E: 2–3 **I:** 20–22 **N:** 0 **B:** 1

Summer

Winter

Black Skimmer
Rynchops niger Length: 18" Wingspan: 44"

The Black Skimmer is a boldly colored bird with a spectacular method of feeding. It flies just above the water surface with its beak open and its longer lower mandible skimming through the water. When it hits a fish, the bird snaps its bill shut to capture the prey. Adding to the excitement is the speed with which the birds skim across the water in large groups. When not feeding, Skimmers rest on beaches or often fly about in flocks of hundreds of birds in long undulating lines, roller-coaster fashion. Breeds coastally on beaches and sandbars with sparse vegetation. Nest is a scrape in the sand. **Voice:** Short barking sounds like *vark vark*.

Winter

Breeding Info — **E:** 4–5 **I:** 21–23 **N:** 0 **B:** 1

Summer

Atlantic Puffin
Fratercula arctica Length: 13" Wingspan: 21"

The Atlantic Puffin, sometimes called the parrot of the sea, is one of America's most charismatic birds, with its sturdy little body and colorful bill that can carry over 20 fish at a time. It lives in the North Atlantic from Maine to Greenland to northern coastal Europe. Its southernmost breeding ground is on the rocky islands off the coast of northern Maine, a territory it shares with Razorbills and Murres. It nests in underground burrows between the crevices of boulders and flies out to the ocean to catch fish and bring them back to feed its young. Outside the breeding season, the birds live at sea. **Voice:** Sounds like a low-pitched chainsaw, *rrrrrrah raar.*

Summer

Breeding Info — **E**: 1 **I**: 35–45 **N**: 43–55 **B**: 1

LAND BIRDS

Turkey Vulture
Cathartes aura Length: 26" Wingspan: 67"

Widespread and common over all of the lower
48 states, the Turkey Vulture may seem a bit
ominous since it is attracted to and feeds on
dead animals, but these birds are harmless. It is
one of the few birds with a well-developed sense
of smell, which helps it find food. Adult has a
blackish body, brownish wings, and a reddish
featherless head; juvenile is similar but with a
blackish head for the first year. Turkey Vultures
are more commonly seen flying than perched,
with their black wings with a silver trailing edge
held in a shallow V and the bird always tilting
from side to side. Nests in hollow trees, caves,
rock crevices; no nest built. **Voice:** Generally
silent, but may hiss.

Breeding Info — **E**: 1–3 **I**: 38–41 **N**: 70–80 **B**: 1

Black Vulture
Coragyps atratus Length: 25" Wingspan: 59"

The Black Vulture is most common in the southern states, but is increasing its range into the Northeast and Midwest. Its typical and distinctive flight pattern is short glides on flat wings followed by a burst of "frantic" flaps. It soars with wings flat or in a slight dihedral. In flight the Black Vulture is all black except for pale wing tips (visible only in flight) and gray featherless head. It feeds mostly on carrion that it finds by sight. Nests in hollow trees, caves, abandoned buildings; no nest is built. **Voice:** Generally silent.

Breeding Info — **E**: 1–3 **I**: 37–48 **N**: 70 **B**: 1

Osprey

Pandion haliaetus Length: 23" Wingspan: 63"

The Osprey is certainly our hawk most closely associated with water, for its diet consists almost exclusively of fish. Expect to see it along the coasts or near large lakes and rivers. It typically hovers high above the water and then dives to catch fish in its talons. When flying, it always turns the caught fish to face forward. Ospreys are largely black and white, and have a white head with a black mask; in flight their wings are slightly bent, creating an M shape in silhouette. They often vocalize around the nest area. The nest is large, composed of sticks, lined with water weeds or seaweed, and built on dead trees, power-line poles, or human-made platforms.
Voice: A rising and descending whistled *ee ee ee ee ee* and a sharply rising *waaaeeh*.

Breeding Info—E: 2–4 **I:** 34–40 **N:** 49–56 **B:** 1

Swallow-tailed Kite

Elanoides forficatus Length: 22" Wingspan: 51"

The long forked tail, streamlined body, long pointed wings, and formal black-and-white plumage make the Swallow-tailed Kite one of North America's most elegant birds. Add to this its buoyant and effortless flight as it chases after aerial insects and you have a bird that is hard to beat. It lives in open areas, woodlands, and swamps and is generally limited to Florida and nearby states for breeding, then leaves for South America in winter. Nest a platform of twigs and sticks lined with Spanish moss and placed in tree-tops 60–130 ft. high. **Voice:** Very high-pitched *ee ee ee.*

Breeding Info — E: 2–4 **I**: 28 **N**: 36–42 **B**: 1

2nd year

Golden Eagle

Aquila chrysaetos Length: 30" Wingspan: 79"

The Golden Eagle is mostly found in the varied open habitats of western North America, such as mountains, foothills, deserts, and grasslands, where it feeds on mammals like ground squirrels and jackrabbits. In flight the adult is a dark brown bird with a golden nape and very long broad wings held in a shallow V; its head projects less in front of the wings than its tail does behind. Immature birds are similar to adults but with broad white band at base of tail and variable white patches at base of primaries. Large nest of sticks is most often placed on a cliff ledge. **Voice:** Generally quiet but may give a throaty, low-pitched *kwow kwow kwow*.

Immature

Breeding Info—**E**: 1–4 **I**: 43–45 **N**: 72–84 **B**: 1

Adults, female left
and male right

Bald Eagle

Haliaeetus leucocephalus Length: 33"
Wingspan: 80"

This is the imposing symbol of the United States
of America, generally seen along coasts and large
rivers or lakes, where it hunts for fish, ducks,
coots, mammals, and carrion, depending on
availability. Its familiar white head and tail are not
gotten until the bird is 4 to 5 years old; before
that, the bird is mostly dark with variable white
mottling. Its silhouette in flight is like a large dark
2 x 10 plank with head and tail projecting equally
in front and in back. Female larger than male.
Eagles build a large nest of sticks and branches
near the top of dead trees and this may be
reused. **Voice:** Calls are high-pitched and squeaky
and seem out of place for such a grand bird.

Juvenile

Breeding Info – **E**: 1–3 **I**: 34–36 **N**: 70–84 **B**: 1

Juveniles

Cooper's Hawk
Accipiter cooperii Length: 16" Wingspan: 31"

The Cooper's Hawk is the slightly larger relative of the Sharp-shinned Hawk and is similar
in appearance and behavior. The Cooper's is
crow-sized with a long rounded tail. The smaller
Sharp-shinned is Blue Jay–sized with a shorter
square-tipped tail. Adult upperparts are gray,
underparts whitish barred with reddish brown;
juvenile upperparts are brown, underparts whitish heavily streaked with brown. The Cooper's is
a widespread breeder across the United States
and southern Canada, and lives in open woodlands, where it nests in forked tree branches.
Nest of sticks lined with bark. **Voice:** A guttural
grating *gakgakgakgak*.

Adult

Breeding Info—E: 3–6 **I:** 32–36 **N:** 27–34 **B:** 1

Juveniles

Sharp-shinned Hawk
Accipiter striatus Length: 11" Wingspan: 23"

Along with the Cooper's, one of the hawks most likely to be seen in your backyard, because it eats birds, including those attracted to feeders. When a Sharp-shinned is in the area, smaller birds often remain motionless and quiet, then resume their activities when it leaves. It is a small hawk (Blue Jay–sized) with a long squared-off tail and short broad wings. The similar-looking Cooper's is larger (crow-sized) and has a longer, more rounded tail. Adults have gray upperparts and barred reddish-brown underparts. Juveniles have brownish upperparts and whitish underparts streaked with brown. Nest is a platform of sticks in dense trees. **Voice:** A high-pitched, harsh _keu keu keu_.

Adult

Breeding Info – E: 3–8 **I:** 30–32 **N:** 21–28 **B:** 1

Center, adult female; others juveniles

Northern Harrier

Circus cyaneus Length: 19" Wingspan: 43"

The Northern Harrier is almost always seen flying low over fields or marshes, its long wings held in a shallow V and the bird tipping from side to side. It courses back and forth, then suddenly drops down as it tries to catch a small bird or mammal off-guard. Harriers have long wings and a long tail with a distinctive white patch at its base. It breeds across much of the upper two-thirds of North America and winters over all of the lower 48 states. The adult male is pale gray above and whitish below, with black wing tips. The female and young are brown above and reddish-brown below. Their nest is made of grasses and placed on the ground in marshes. **Voice:** A scratchy *kree kree kree*.

Adult male

Breeding Info — **E**: 3–9 **I**: 31–32 **N**: 30–35 **B**: 1

Top 2 juveniles;
bottom 2 adults

Red-shouldered Hawk

Buteo lineatus Length: 17" Wingspan: 40"

This is a medium-sized hawk strongly associated with swamps and wet woods, where it feeds on frogs, snakes, lizards, and other amphibians and reptiles. It is especially common in the Southeast, where it is often seen perched in trees or along power lines looking for prey in the wet roadside ditches. The adults have striking black-and-white-checkered wings above and below; juveniles have wings barred with brown and white and a whitish underbody streaked with brown. At all ages, a pale crescent of light shines through the base of their outer wing. Their nest of twigs and sticks is placed in a tree. **Voice:** A wild, drawn-out, descending screech, like *kreeeaaah*.

Breeding Info — **E**: 2–6 **I**: 28 **N**: 35–49 **B**: 1

Adult

Right large bird adult; left large bird juvenile

Broad-winged Hawk

Buteo platypterus Length: 15" Wingspan: 34"

The Broad-winged, during breeding, is a secretive hawk of woodlands, but in fall, as it migrates
to South America, it joins spectacular flocks of
hundreds to thousands. The flocks, called ket-
tles, consist of swirling masses of birds that rise
on thermals of hot air and then, when they run
out of lift, glide south until they find the bottom
of the next thermal. Adults' most distinctive
feature is a dark tail with a dominant wide white
band across the middle; underparts are broadly
barred with reddish brown. Juveniles have
numerous thin pale tail bands and their whitish
underparts are streaked with brown. Nest of
twigs is placed in a forked trunk. **Voice:** A distinctive, very high-pitched, drawn-out whistle
with an introductory note like *tsitseeeeh*.

Adult

Breeding Info—E: 1–4 **I:** 30–38 **N:** 35–40 **B:** 1

Top and right adults; others juveniles

Red-tailed Hawk

Buteo jamaicensis Length: 20" Wingspan: 49"

The Red-tailed Hawk is the most common hawk of open farmlands and grasslands, where it feeds primarily on rodents like voles, but also on some birds and snakes. It is frequently seen perched on dead trees or telephone poles looking for prey along the sides of highways. It is a large hawk with much variation in appearance, but most of the time it can be recognized by its broad reddish-brown tail and mildly streaked belly band. Juvenile has a light brown tail with multiple fine dark bands and a coarsely streaked wide belly band. Red-taileds fly with their wings in a shallow V and sometimes do a stationary hover facing into the wind, called kiting. Bulky nest of twigs is placed in a tree. **Voice:** A high-pitched scratchy *sheeaah*.

Adult

Breeding Info — **E:** 1–5 **I:** 28–35 **N:** 44–46 **B:** 1

Adult

Swainson's Hawk

Buteo swainsoni Length: 19" Wingspan: 51"

A primarily western hawk that lives in prairies, agricultural fields, and other open land, where most of the time it feeds on large insects (crickets, grasshoppers) gathered by perching above and swooping down or while standing on the ground. During breeding feeds more on small mammals, which it gives to its young. Like the Broad-wing in the East, the Swainson's may gather in large flocks while migrating, using thermals for lift. The Swainson's main identification clue is its extensive dark brown bib, which gives it a hooded look. Nest of twigs and finer material placed in groves of trees or shrubs in open habitat. **Voice:** A descending screech like *shreeeeaaah*.

Adult

Breeding Info – E: 2–4 **I:** 28–35 **N:** 30 **B:** 1

Top, male;
bottom, female

American Kestrel

Falco sparverius Length: 9" Wingspan: 22"

Our smallest, most common, and most wide-spread falcon; found in a variety of open habitats from rural to urban. Falcons have long pointed wings and fairly long tails. Adults have two dark vertical bars across a white face. Male has reddish-brown back and bluish-gray wings; female has brown back and wings. Kestrels eat mostly large insects caught off the ground or in the air; these include dragonflies, butterflies, crickets, and grasshoppers. They breed in tree holes, nest boxes, or outbuilding nooks, and finding a suitable site may be a limiting factor on their breeding in some areas. **Voice:** A high-pitched quickly repeated *kleekleekleeklee*.

Breeding Info – **E**: 3–7 **I**: 29–31 **N**: 29–31 **B**: 1

Male

Merlin

Falco columbarius Length: 10" Wingspan: 24"

This is the falcon "in a hurry"; a Merlin no sooner shows up than it is gone, with its direct and speedy flight. It is only slightly larger than the Kestrel but has a shorter tail, broader wings, and only a faint single vertical bar across its face. Underwings and tail are dark; whitish underbody is heavily streaked. The adult male is bluish gray above; adult female and juvenile are dark brown above. It preys on smaller birds and sometimes on insects. Merlins breed in Canada and Alaska but winter in most of the lower 48 states. Can nest on a cliff ledge or on the ground, and usually just forms a scrape rather than building a nest. **Voice:** High-pitched series of flat calls like *kwee kwee kwee.*

Breeding Info—E: 2–7 **I:** 28–32 **N:** 25–35 **B:** 1

Adult

Peregrine Falcon

Falco peregrinus Length: 16" Wingspan: 41"

The Peregrine Falcon is one of North America's best fliers and most skillful hunters. Its flying is characterized by agility, speed, and seeming effortlessness. It has a single broad dark vertical bar across its face. Adults have underwings and body uniformly barred gray except for white chin and breast. Adult male more bluish; female more brownish. Juvenile similar to adult but brownish and with streaking on breast. Peregrines nest in the wild on cliff faces; but occasionally, they also breed in cities, where they nest on building ledges and feed on pigeons. Eggs are usually laid in a simple depression on a cliff ledge. **Voice:** A loud high-pitched *shreeshreeshreeshree*.

Adult

Breeding Info — **E**: 3–4 **I**: 28–33 **N**: 30–42 **B**: 1

Male

Ring-necked Pheasant
Phasianus colchicus Length: male 33";
female 21" Wingspan: 31"

Best known for its long thin tail and the striking
plumage of the male, which has an iridescent
green head, bright red face, and white neck-
ring; his body is coppery to iridescent green.
The female is camouflaged by her buffy body
with brown spotting. Ring-necked Pheasants
vary worldwide, and many subspecies have been
introduced into North America for game hunters, so you may see many variations in plumage.
They feed and breed in farmlands, hedgerows,
and marsh edges. Nest of grasses and leaves
placed on ground. **Voice:** Male call is a loud
skwagock often answered by female's *kia kia*; also
tukatuk tukatut.

Female

Breeding Info—**E**: 10–12 **I**: 24 **N**: 0 **B**: 1

Ruffed Grouse

Bonasa umbellus Length: 17" Wingspan: 22"

An inconspicuous chickenlike bird of woods
and fields that you are more likely to hear than
see, if you know what to listen for. The male
in spring stands on logs or elevated spots and
makes an increasingly rapid, low-pitched series
of throbbing sounds by cupping his wings
against his body. Back, head, and tail are either
reddish-brown or grayish-brown; flanks are
whitish with bold black barring; tail is long and
barred; and crest can be raised into a triangular
peak when alert. Feed on seeds, insects, and
tree buds. Nest of grasses and leaves placed on
ground. **Voice:** Calls include a quiet *gweek, gweek,
gweek.*

Breeding Info — **E**: 6–15 **I**: 21–28 **N**: 0 **B**: 1

Male

Wild Turkey

Meleagris gallopavo Length: male 46";
female 37" Wingspan: male 64"; female 50"

A magnificent large bird seen in flocks of 5 to 25
feeding in fields or along roadsides. Becoming
increasingly common due to reintroductions
across the United States. Feeds by scratching
on ground with strong feet to get at vegetation,
seeds, fruits, and insects. The male is larger
than the female, has a spur on his lower leg, and
does the spectacular tail-fanning and gobbling
courtship display. When airborne, they look
shockingly large. Male keeps a harem of several
females who go off on their own and lay eggs
after mating. Nest of grasses and leaves placed
on ground. **Voice:** Male gobbles and gives *kwuk*s
and rattles.

Breeding Info—E: 6–20 **I:** 27–28 **N:** 0 **B:** 1

Male

California Quail

Callipepla californica Length: 10" Wingspan: 14"

An endearing quail because of its cute topknot
feathers and habit of staying in small groups
called coveys. They inhabit brushy areas in open
woodlands, field edges, roadsides, and suburbs;
come to feeders with seed on the ground. Both
sexes have a pale yellow belly with strongly
scaled look. The male has a black topknot, black
face with white streaks, and dark patch on the
belly. The female has a brown face with no white
streaks and no dark belly patch. Nest of grasses
and leaves usually placed on ground. **Voice:** Call
sounds like *chi ca go*; also gives a spitting *pit pit pit*.

Breeding Info—E: 12–16 **I:** 18–23 **N:** 0 **B:** 1

Female

Male

Gambel's Quail

Callipepla gambelii Length: 11" Wingspan: 14"

This is the southwestern counterpart to the California Quail, which lives mostly in the West Coast states. The two species have similar behavior and plumage, except that the Gambel's Quail has a yellowish belly with no scaling. The male has a black topknot, black face with white streaks, and dark patch on the belly. The female has a brown face with no white streaks and no dark belly patch. They favor arid shrubby areas and riparian habitats. They feed on seeds, buds, and berries, and will come to seed on the ground at bird feeders. Nest of grasses and leaves is on the ground. **Voice:** A muted *ga way gaga, ga way gaga* and a spitting *pit pit pit.*

Breeding Info — **E:** 9–14 **I:** 21–24 **N:** 0 **B:** 1

Female

Male

Northern Bobwhite
Colinus virginianus Length: 9¾" Wingspan: 13"

Known best for its loud ringing call of "Bob . . .
White" given by the male in spring and summer,
usually as he perches on an elevated spot. This is
a very rounded, short-tailed, short-necked bird
with a short stubby bill. It is much more likely to
be heard than seen as it moves about in brushy
fields and open woods in small groups called
coveys. The male's eyebrow and throat are white,
while these areas on the female are buffy. They
feed on seeds, leaves, and some insects. The nest
is a depression in the ground filled with grasses
and leaves, located under vegetative cover.
Voice: Calls include *bob white*, a loud *koilee koilee*,
and a soft *tu tu tu*.

Breeding Info—E: 12–14 **I**: 23 **N**: 0 **B**: 1

Male

Band-tailed Pigeon

Patagioenas fasciata Length: 14½" Wingspan: 26"

This is a large pigeon of the West Coast and Southwest mountains, where it travels in flocks. It feeds much of the time in trees, eating fruits and seeds, but also comes to the ground to feed on grains. It will visit bird feeders that are traylike and have a large landing platform. It is named for the broad pale gray band covering the tip of the tail, often seen in flight. On the perched bird, look for the thin white half collar on the back of its neck with extensive iridescent speckling below. The bill is yellow with a dark tip. Nest is a flimsy collection of twigs placed on a horizontal branch of a tree, often an evergreen. **Voice:** A low-pitched cooing like *hooah wooo*.

Breeding Info — **E:** 1–2 **I:** 18–20 **N:** 25–27 **B:** 2–3

Eurasian Collared-Dove

Streptopelia decaocto Length: 13" Wingspan: 22"

The real story behind this bird is the incredible expansion of its range. It was originally from the Middle East and then spread into Europe. Introduced into the Bahamas in 1974, it spread from there to Florida. It is now breeding in over two-thirds of the United States and has made appearances in almost every state. It is a beautiful pale gray with a thin black half collar and a broad white band on the end of its tail. **Voice:** Repeated cooing of *cucu cooo*, and a kazoolike buzz in display flight.

Breeding Info — **E**: 2 **I**: 14–16 **N**: 15–18 **B**: 2–5

Rock Pigeon

Columba livia Length: 12½" Wingspan: 28"

This is the familiar pigeon, often dismissed due
to its prevalence. But it should not be over-
looked, for it is one of our most skillful avian
fliers, and its tameness offers us a chance to see
pigeon behavior close-up. It was introduced to
North America in the early 1600s. Well known
for its homing abilities, it has been used for
centuries to carry messages. Pigeons' success
is in part due to their tolerance of humans and
their acceptance of nest sites on building ledges,
which resemble the cliffs in their native habitats.
The nest is a loose collection of twigs placed on a
ledge. **Voice:** A low-pitched gurgling *cucucurooo*.

Breeding Info—E: 1–2 **I:** 18 **N:** 25–29 **B:** 2–5

White-winged Dove

Zenaida asiatica Length: 11½" Wingspan: 19"

A dove primarily of the Gulf states and the Southwest, it is at first glance a Mourning Dove look-alike. But the lack of dark ovals on its wings and the bold slash of white across the edge of its folded wings, which can also be seen in flight, clearly distinguish this handsome dove. In the Southwest it is a resident of arid areas, while in the Gulf states it is common in landscaped suburbs and parklands. It regularly visits feeders with large trays or seed sprinkled on the ground. Saucerlike nest of sticks and weed stems placed on a horizontal branch or atop another bird's previous nest. **Voice:** *Coocoo cuh coooo.*

Breeding Info — **E:** 1–4 **I:** 13–14 **N:** 13–16 **B:** 2–3

Mourning Dove

Zenaida macroura Length: 12" Wingspan: 18"

There is practically nowhere you can go in the lower 48 states and southern Canada without encountering Mourning Doves. Not only are they geographically widespread, they are habitat diverse, being found in deserts, woods, parks, and suburbs. They are named for their mournful drawn-out cooing. They are our only medium-sized dove with a sharply pointed tail and black oval markings on their wings; these help distinguish them from the similar White-winged Dove. Their nest is a see-through arrangement of twigs, male supplied and female built. **Voice:** Low-pitched *hoowaoo hoo hoo hoo.*

Breeding Info— **E**: 2 **I**: 14–15 **N**: 12–14 **B**: 2–3

Male

Female

Common Ground-Dove

Columbina passerina Length: 6½" Wingspan: 10½"

These are the cutest pocket-sized doves you will
ever see. They spend most of their time walking
about sandy areas near taller vegetation. They
are very small, deep-chested, and short-tailed,
with a short neck and relatively large head.
The male has a rosy breast with bluish-black
iridescent spots on his wings; the female is more
evenly sandy to brownish with dark brown
to coppery brown spots on her wings. Their
underwings are a bright rusty red. Their nest is
made of sticks, rootlets, and grass placed on the
ground or up in shrubs or trees. **Voice:** A low-
pitched upslurred *ooowah ooowah*.

Breeding Info — E: 2–3 I: 12–14 N: 11 B: 2–4

Greater Roadrunner

Geococcyx californianus Length: 23"
Wingspan: 22"

Hard to think of the roadrunner without thinking of the familiar cartoon character Wile E.
Coyote and his hopeless attempts to catch it
with mail-order gadgets. Real roadrunners spend
more time chasing lizards and other small desert
animals. They may also eat fruit, insects, snakes,
and the eggs or young of other birds. Running is
their modus operandi, and they can move along
at a 15 mph clip or faster for shorter dashes.
They prefer arid areas with sparse brush, and
place their nest of sticks and rootlets in a shrub
or cactus 3–15 ft. high. **Voice:** Deep, mellow,
well-spaced cooing trailing off at the end, *wooh
wooh wooh whoa whoa.*

Breeding Info — E: 4–6 **I:** 20 **N:** 17–19 **B:** 1–2

Barn Owl

Tyto alba Length: 16" Wingspan: 42"

Although called a "barn" owl, this species nests
in a wide variety of locations besides barns,
including outbuildings, church steeples, tree
hollows, woodchuck burrows, holes in banks and
cliffs, and even nest boxes if they have the right
dimensions. Its white, heart-shaped facial disk
and tawny colors are unmistakable. With the
most acutely developed sense of hearing of all
our owls, it can pinpoint prey with 100 percent
accuracy in total darkness. It hunts over a wide
variety of habitats, including deserts, open country, grasslands, marshes, farmlands, suburbs, and
cities. Eats mostly mice and rats, also insects,
bats, and reptiles. **Voice:** A distinctive harsh raspy
screech unlike that of other owls, like *sshrreet*.

Breeding Info—**E**: 4–7 **I**: 32–34 **N**: 45–58 **B**: 1–2

Brown morph

Red morph

Eastern Screech-Owl

Megascops asio

Length: 8½" Wingspan: 20"

Screech-Owls are our most common and wide-spread small owls, nesting primarily in cavities such as those in old trees, woodpecker holes, and birdhouses. They are our only small owls with yellow eyes and ear tufts; the Eastern's bill is pale. They become active at dusk, which is when they are most frequently heard and when you can see the male bring food to the female during incubation. Screech-Owls eat mice, insects, reptiles, and amphibians. **Voice:** A descending screech and a monotone call of repeated notes.

Red morph

Breeding Info — **E:** 3–5 **I:** 21–28 **N:** 30–32 **B:** 1

Gray morph

Western Screech-Owl
Megascops kennicottii

Length: 8½" Wingspan: 20"

The Western Screech-Owl is nearly identical to
the Eastern Screech-Owl, with the exception
of its dark bill. Sometimes the best way to distinguish between them is by location: if you've
spotted one within the area of the Western range
map, for example, chances are it's a Western.
Screech-Owls are polychromatic and occur in
red, gray, and intermediate brown morphs. The
Western Screech-Owl is mostly gray, with only
about 10 percent red morphs. Red morphs are
more common in the Eastern Screech-Owl, especially in the Southeast. See Eastern Screech-Owl
for more information. **Voice:** A series of repeated
high-pitched hoots speeding up at the end.

Gray morph

Breeding Info – **E**: 3–5 **I**: 21–28 **N**: 30–32 **B**: 1

Adult

Great Horned Owl

Bubo virginianus Length: 22" Wingspan: 44"

The Great Horned Owl is one of our largest and fiercest avian predators and the most widespread of all of our owls, living throughout the continental United States and most of Canada. The "horns" are not horns or ears, which are lower and within the facial disk; they are just tufts. The Great Horned can feed on mammals as large as skunks and even birds as large as hawks. It does not build a nest of its own but usurps those of herons, squirrels, and hawks. At 6 to 8 weeks, young owls leave the nest and perch on nearby branches before taking their first flights. **Voice:** Deep low-pitched hoots in 1s, 2s, or 3s, like *hoohoohoo hoohoo hoo*.

Juveniles

Breeding Info — **E**: 1–4 **I**: 28–30 **N**: 35 **B**: 1

Snowy Owl
Bubo scandiacus Length: 23" Wingspan: 52"

Your best chance of seeing a Snowy Owl is in
winter, when they migrate to southern Canada
and the northern states from their breeding
grounds in the far north. The owls can be irrup-
tive, with larger numbers moving farther south
in times of population boom or food scarcity;
most of these are 1st-year birds. Snowy Owls
favor open country such as tundra, barren fields,
short-grass areas, and coastal dunes, where they
perch on the ground, a post, or a building. They
feed day or night on rabbits, fish, waterfowl, and
carrion. Adults are the whitest, males more than
females. Nest a shallow scrape on a mound lined
with feathers and moss. **Voice:** Very high-pitched
finchlike *tseeah* and a low-pitched croaking *kwohk*.

Breeding Info — **E**: 3–10 **I**: 32–39 **N**: 14–28 **B**: 1

Adult

Barred Owl

Strix varia Length: 21" Wingspan: 42"

This is the owl most likely to be heard hooting during the day, its call sounding much like the phrase "Who cooks for you?" It is also our only common large owl with dark eyes (its cousin, the Spotted Owl, is uncommon), for all other large owls have yellow eyes. It can also be recognized by the barring on the chest with thin streaking below. Barred Owls eat rodents, rabbits, amphibians, and reptiles, caught at night. They typically nest in woods or wooded swamps, using a large cavity in a tree or an abandoned hawk or crow nest. **Voice:** *Hoo hoo hoohooo;* also a downslurred scream.

Juvenile

Breeding Info—**E**: 2–4 **I**: 28–33 **N**: 40–45 **B**: 1

Adult

Adult

Burrowing Owl

Athene cunicularia   Length: 9½" Wingspan: 21"

Burrowing Owls breed throughout the West and
in Florida. A personable owl that can be active by
day or night and lives in burrows from which one
or more may poke its head out to look at you. In
Florida they are not bothered by humans and will
dig their burrows right in people's lawns. They
use their long legs to kick the dirt out of the burrow during excavation. In the West they may use
old prairie dog holes. They eat insects and small
mammals. Seeing a set of nestling Burrowing
Owls at the burrow entrance is very adorable.
Voice: Midpitched dovelike *kwacooo*, also a high-
pitched *cheecheechee*.

Breeding Info — **E**: 6–11 **I**: 21–28 **N**: 28 **B**: 1

Juveniles

Female

Male

Common Nighthawk

Chordeiles minor Length: 10" Wingspan: 24"

The buzzy call of the Common Nighthawk may be what first alerts you to their presence. Their long pointed wings with white bars near the tip (larger on males) and their snappy wingbeats make them easy to identify. You can see their darting flight path in the late afternoon as they start to feed on aerial insects. In fall they migrate in the early evening, often in groups of many to a thousand birds, alternating feeding with straight-line movement south. Nest is a scraped depression on barren ground or even a gravel rooftop. Males do a diving flight display accompanied by a booming sound created by their wings. **Voice:** A buzzy *beeent*.

Male

Breeding Info—**E**: 2 **I**: 19–20 **N**: 21 **B**: 1

Male

Female

Ruby-throated Hummingbird

Archilochus colubris Length: 3½" Wingspan: 4½"

The major hummingbird species of the East.
Many other species can show up in the East,
but they are relatively rare. The male has a black
mask and ruby-red iridescent throat that can
appear all dark from certain angles or when not
lit by sunshine; he has an iridescent green back
and dingy "vest" on an otherwise pale belly. The
female has an iridescent green back and white
underparts with a touch of green or buff on the
flanks. They eat insects, nectar, pollen, spiders,
sap from sapsucker holes, and come to hummingbird feeders. Nest of plant down and spider
silk covered with lichens, placed on a small horizontal limb. **Voice:** Song is a long series of *chits*.

Male

Breeding Info — **E**: 2 **I**: 16 **N**: 30 **B**: 1

Male

Black-chinned Hummingbird

Archilochus alexandri Length: 3½" Wingspan: 4¾"

One of the most widespread hummingbirds of
the West, living in a great variety of habitats
from wild canyons and streamsides to suburban
backyards. The male has a distinctive black chin
bordered by a violet throat. The female's upperparts are iridescent green above; her underparts
are white with a touch of green or buff on the
flanks. Very similar to the female Ruby-throated,
but their ranges, for the most part, do not over-
lap. Eats nectar, insects, spiders, and comes to
hummingbird feeders. Nest of plant down collected from willows and underside of sycamore
leaves bound with spider silk, placed in tree or
shrub. **Voice:** A rapid *ch'ch'ch'ch'chit*.

Female

Breeding Info – **E**: 1–3 **I**: 13–16 **N**: 21 **B**: 1–2

Male

Anna's Hummingbird

Calypte anna Length: 3¾" Wingspan: 5¼"

Not only is the Anna's the most common hummingbird along the West Coast, it also is a year-round resident, treating people to its beauty in all 12 months. The male is distinctive, with a full helmet of reddish-orange iridescence. The female appears stocky, with a relatively short bill, shorter than her head length. She is iridescent green above and on her flanks; she has a variable small rosy-pink spot in the center of her throat. Eats nectar and insects. Nest of downy plant fibers bound together with spider silk and coated with lichens is placed in shrubs. **Voice:** Male song a long series of high-pitched sounds; female calls 1- to 2-syllable harsh *chit*s or *chidit*s.

Female

Breeding Info — **E**: 2 **I**: 14–19 **N**: 18–20 **B**: 1–2

Male

Broad-tailed Hummingbird

Selasphorus platycercus Length: 3¾" Wingspan: 5¼"

The most common hummingbird in the mountains of the West, preferring to nest in higher-elevation meadows and shrub thickets up to about 10,000 ft. Male makes a distinctive repeated, short, metallic trill created by his outer wing feathers when flying (not hovering), which can alert you to his presence. He looks similar to a male Ruby-throated but is gray (rather than black) behind the eye. The female is iridescent green above with buffy flanks and only faintly spotted throat; her tail is noticeably longer than her wings on perched bird. Eat insects, spiders, nectar, and tree sap. Nest of plant down and spider silk placed on horizontal limb of tree or shrub. **Voice:** Short *tsip*, multisyllable *chitterdit*, and drawn-out *dzeeet*.

Female

Breeding Info—E: 2 **I**: 16–17 **N**: 18–23 **B**: 1–2

Male

Rufous Hummingbird

Selasphorus rufus Length: 3½" Wingspan: 4½"

The Rufous Hummingbird regularly breeds farther north in the West than any other hummingbird, extending well up the southern Alaska coastal regions. Male upperparts and flanks mostly orangish-brown with some variable green flecking on back and crown; throat iridescent orange-red and slightly flared at corners. Differs from similar male Allen's in having little or no green on upperparts. Female Rufous nearly identical in the field to female Allen's and distinguishable only by shapes of tail feathers and of course by breeding range. Eat insects, spiders, and tree sap. Nest of downy plant fibers and moss bound by spider silk placed on a sloping branch of shrub or tree. **Voice:** A high trill made by male wings in flight display. High-pitched, short, harsh *chit chidit chit*.

Female

Breeding Info — **E**: 2 **I**: 12–14 **N**: 20 **B**: 1–2

Male

Allen's Hummingbird

Selasphorus sasin Length: 3½" Wingspan: 4¼"

During breeding, the Allen's is the "orange hummingbird" of the California coast, since the look-alike Rufous Hummingbird breeds mostly from Oregon north. Allen's live in a variety of habitats, from woods to thickets and parks to backyards. The male is orangish-brown with substantial iridescent green on his back and crown. The female is virtually identical to the female Rufous Hummingbird, except for range during breeding. Feed on nectar, insects, spiders, and at hummingbird feeders. Nest of moss, dried weed stems, and willow down bound with spider silk usually straddles a branch of shrub or tree. **Voice:** A high trill made by male wings. High-pitched, short, harsh *chit chidit chit*.

Male

Breeding Info — **E**: 2 **I**: 15–17 **N**: 22–25 **B**: 1–2

Chimney Swift

Chaetura pelagica Length: 5¼" Wingspan: 14"

In the East, hot summer days in small towns go together with the chittering calls and stiff fluttery flight of Chimney Swifts as they circle above the houses where they nest in chimneys. Their short bodies and thin sickle-shaped wings have prompted some to describe them as a "cigar with wings." They are dark grayish-brown overall with a paler gray chin and throat; the sexes look alike. They break twigs off trees in flight and cement them to the inner walls of chimneys with a sticky salivalike substance to build their nests. After breeding, they leave the United States in winter. Eat aerial insects. **Voice:** High-pitched chittering and high-pitched repeated *tseer tseer tseer.*

Breeding Info — **E**: 4–5 **I**: 19 **N**: 14–18 **B**: 1

White-throated Swift
Aeronautes saxatalis Length: 6½" Wingspan: 15"

This is the common swift of the interior West
and the southern coast of California. It is most
likely seen near its nesting habitat of inland or
coastal cliffs, where it builds its cuplike nest of
feathers and plant material held together with
a salivalike glue in the cliff crevices. The striking
white patches on the throat and hip are unique
among our other regularly occurring swifts. This
species overwinters in the extreme Southwest
and is our only swift in winter, since all other
species winter south of the United States. Eats
aerial insects. **Voice:** Rapid series of high-pitched
tsetsetsetse.

Breeding Info – **E**: 3–6 **I**: 24 **N**: 40–46 **B**: 1

Male

Belted Kingfisher

Megaceryle alcyon Length: 13" Wingspan: 20"

If you are anywhere near water and hear a strident rattling call, scan for a Belted Kingfisher.
They are noisy active birds often seen hovering
over the water and then diving down to catch
fish; alternately they may be sitting on the perfect perch overlooking water, ready to hunt or
chase another Kingfisher. The male has a single
blue breastband; the female additionally has a
brown belly band. Both have a white patch on
the outer wing that is prominent in flight and
a huge bill and bushy crest that gives the birds
a rakish look. They eat fish, amphibians, crustaceans, and insects. Their nest is an excavated
tunnel in an earthen bank. **Voice:** Very rapid
rattling *chchchch*.

Female

Breeding Info – E: 5–7 **I**: 22–26 **N**: 18–28 **B**: 1

Red-headed Woodpecker

Melanerpes erythrocephalus Length: 9¼"
Wingspan: 17"

Wow, what a beauty! No intricate patterning here, just simple bold blocks of three colors—red, black, and white—creating one of our most spectacularly colored birds. The Red-headed prefers open country with scattered trees, including farmlands, orchards, open woodlands, and suburbs. Male and female look alike, but it takes them 2 years to achieve the adult plumage. Before that they gradually develop the red head, and they have dark barring across the white in their wings. They eat insects, fruit, bird eggs, and sap from trees; come to feeders for suet. Nest excavated in live or dead tree; may use birdhouses. **Voice:** A slightly rising *queeek* and a raspy *churrrr*.

Breeding Info — **E**: 4–5 **I**: 12–13 **N**: 30 **B**: 1–2

Male

Female

Red-bellied Woodpecker

Melanerpes carolinus Length: 9¼" Wingspan: 16"

The Red-bellied is most common in the South-east — boisterous and active, calling and chasing, hitching around trees, and drumming on house gutters or other resonant surfaces. The faint red on the belly is on the extreme lower belly and is practically hidden. The visible red is on the head: for the female, on the nape; for the male, on the nape and crown. It is one of the few woodpeckers that readily breeds in birdhouses; otherwise it excavates its own nest hole. It eats insects, fruit, and seeds; at feeders, orange halves, seed, and suet. **Voice:** A bubbling *churrrr* and a repeated *chew chew chew*.

Male

Breeding Info — E: 3–8 **I**: 12–14 **N**: 25–30 **B**: 2–3

Adult

Red-breasted Sapsucker

Sphyrapicus ruber Length: 8½" Wingspan: 16"

This is the common sapsucker of the Pacific Coast. Unlike other sapsuckers, the sexes look alike. The adults have a deep red head and breast, a mottled black-and-white belly, and a bold white wing patch. A northern subspecies has a light yellow belly. Excavates small holes in trees and then continues to revisit them to drink the accumulated sap. Nest is excavated in trees whose heartwood is softened by fungus. **Voice:** High-pitched downslurred *cheeer*.

Breeding Info — **E**: 3–6 **I**: 12–13 **N**: 25–29 **B**: 1

Juvenile

Female

Juvenile

Yellow-bellied Sapsucker
Sphyrapicus varius Length: 8½" Wingspan: 16"

The only sapsucker in the eastern half of the United States and Canada. Like all our sapsuckers, it excavates small holes in a circular pattern around a tree trunk. When it finds a good flow of sap, it drills a vertical line of holes to exploit it. It then revisits to drink the sap and keep it running. Some hummingbirds visit sapsucker holes to share in this bounty. The drumming of the Yellow-bellied is irregular and suggestive of Morse code. The Yellow-bellied can be recognized by its red forehead and bold white wing patch. The male also has a red chin; the female does not.
Voice: A harsh and shrieking *shreech shreech*.

Male

Breeding Info — E: 5–6 **I:** 12–13 **N:** 25–29 **B:** 1

Male

Female

Hairy Woodpecker

Picoides villosus Length: 9¼" Wingspan: 15"

The Hairy is a larger look-alike of the Downy Woodpecker. One way to distinguish them is by proportions: the Hairy's bill is more than half the length of its head; the Downy's bill is less than half the length of its head. The male has red on his nape; the female does not. Young males in mid- to late summer have reddish feathers on their central crowns. Hairy Woodpeckers can excavate nest holes in very hard woods such as oaks and maples. Eat wood-boring insects.
Voice: Sharp *teeek*; repeated *wikiwikiwiki*.

Breeding Info — **E**: 4–6 **I**: 11–12 **N**: 28–30 **B**: 1

Juvenile

Male

Female

Downy Woodpecker
Picoides pubescens Length: 6¾" Wingspan: 12"

The Downy is our smallest woodpecker and a look-alike with the larger Hairy Woodpecker, except that it has a proportionately smaller bill (less than half the length of its head) and generally small black bars on its white outer tail feathers. The female has no red on her head, while the male has a red nape patch. Their small size means that Downys generally excavate nest holes in softer or partly rotted wood. They eat a variety of insects and come to feeders for seed and suet. **Voice:** A loud *teeek*, a descending whinny, and repeated 3 to 5 *kweeek*s.

Breeding Info — E: 4–5 **I:** 12 **N:** 21 **B:** 1–2

Male

Male

Female

Pileated Woodpecker

Dryocopus pileatus Length: 16½" Wingspan: 29"

This largest North American woodpecker is about the size of an American Crow. It can be recognized by its large size and big bright red crest. Males have a red line off the base of the bill; females do not. Pileateds have the ability to excavate huge gashes in trees as they search for carpenter ants; they can also forage for fruits and insects in a variety of ways, even on the ground in the South. They may come to suet at feeders. They excavate a big nesting cavity with an entrance 3½" in diameter. Their drumming is low-pitched, soft, and trails off at the end. **Voice:** Loud *kekekeke*.

Breeding Info — **E:** 3–5 **I:** 15–16 **N:** 28–32 **B:** 1

Female

Female (Yellow-shafted)

Male (Yellow-shafted)

Northern Flicker

Colaptes auratus Length: 12½" Wingspan: 20"

Northern Flickers keep a surprise for birdwatchers under their wings — a stunning flash of red or yellow seen on the underwing and undertail of the bird as it flies. There are two groups of Northerns: those with red underwings (mostly western), and those with yellow (mostly eastern). The males have black or red "moustaches"; females do not. Flickers often feed on ants on the ground; they also eat fruit and other insects, and at feeders take seed and suet. They excavate their own nests in soft wood and will accept a birdhouse. **Voice:** Loud single *keeoh*, a soft *woika woika*, and a loud *kekekeke*.

Breeding Info — **E**: 7–9 **I**: 11–12 **N**: 14–21 **B**: 1–2

Male (Red-shafted)

Black Phoebe
Sayornis nigricans Length: 7" Wingspan: 11"

The Black Phoebe is an unmistakable bird of the Southwest and California, recognized by its overall black color with a white belly and the continual bobbing of its tail. It is a denizen of water edges, such as shady streams, roadside ditches, and farm ponds. Here it finds an open perch and sallies out to catch aerial insects. Its nest is made of mud and grasses, and is usually stuck to a vertical surface with some overhead protection, such as a cliff, bridge, or wall. Male and female look alike. **Voice:** 2 high-pitched, repeated whistled phrases, like *tsitsee tsitseew*, the last part downslurred.

Breeding Info — E: 4–5 **I**: 15–18 **N**: 14–21 **B**: 1–2

Eastern Phoebe

Sayornis phoebe Length: 7" Wingspan: 10½"

This is the bird that most often nests on houses
and outbuildings, usually under overhead cover.
It often chooses a lamp fixture on a front porch.
The nest is made of mud and strands of grass or
moss, and the birds may reuse it, after refurbish-
ing it, for several years. The Eastern Phoebe is
brownish gray above and pale below, continually
bobs its tail, and during breeding repeatedly
gives a scratchy rendition of *feebee* — not to be
confused with the clear two-note whistle of the
Black-capped Chickadee. It chooses prominent
open perches from which it flies out to catch
aerial insects. **Voice:** A scratchy *feebee* and a
sweet *chirp*.

Breeding Info — **E**: 4–5 **I**: 16 **N**: 18 **B**: 1–3

Say's Phoebe

Sayornis saya Length: 7½" Wingspan: 13"

The Say's Phoebe lives throughout most of the
West and is more likely to be found in the drier
habitats of sagebrush and prairies than our other
two phoebes. It looks like the Eastern Phoebe
but has a rusty belly, and it flares rather than
bobs its tail. It flies out from perches to catch
aerial insects; also hovers over ground to find
other insects. Nest is a platform of grasses,
moss, and cocoons, with little or no mud, placed
in a natural sheltered nook such as a cliff crevice,
a building, or under a bridge. **Voice:** A plaintive
downslurred *peeeuu* and 2 whistled phrases, the
first clear and downslurred, the second buzzy
and rising.

Breeding Info – **E:** 4–5 **I:** 14 **N:** 14–16 **B:** 1–3

Great Crested Flycatcher

Myiarchus crinitus Length: 8" Wingspan: 13"

You are more likely to hear the loud calls of this bird before you see it, for it tends to remain up in the leafy canopy of trees. It is a fairly large flycatcher with a yellow belly, gray breast, and reddish-brown wings and tail. It has a prominent bushy crest and is our only crested flycatcher in the East. It feeds on aerial insects and occasionally berries. It makes its nest in a natural cavity, old woodpecker nest, or birdhouse, usually adding grass, fur, other plant material, and sometimes bits of shed snakeskin. **Voice:** A loud upslurred *weeep* or *kaweeep*, a loud *kaweet weeer*, and a short *prrrt*.

Breeding Info—**E**: 5–6 **I**: 12–15 **N**: 14–21 **B**: 1

Western Kingbird
Tyrannus verticalis Length: 8¾" Wingspan: 15½"

This is the most common and widespread king-bird in the West. It can be recognized by the pale gray breast that blends into its yellow belly, and its blackish tail with a dark tip and white edges. Like all kingbirds, it loves open perches, especially fence posts, from which it can fly out to catch aerial insects or drop down to the ground to get crawling insects. It prefers open farmland, streamsides, arid scrub, and suburbs. Bulky nest of twigs, plant stems, and rootlets is placed on a horizontal limb, 8–40 ft. high. **Voice:** A harsh *kit kit kit* and a more extended *kit kit kidit kadeet kitterdit*.

Breeding Info—E: 3–5 **I:** 13–15 **N:** 16–18 **B:** 1–2

Eastern Kingbird
Tyrannus tyrannus Length: 9" Wingspan: 15"

The Eastern Kingbird is ready, at the drop of a hat, to chase any other bird out of its territory and away from its nest, whether the intruder is as small as a Red-winged Blackbird or as large as a Red-tailed Hawk, often repeatedly striking the other bird from above. It is easy to see why its scientific name is *Tyrannus tyrannus* — it's too (two?) tyrannical. This bird is black above, white below, and has a white band at the tip of its tail. Its nest is a somewhat disheveled collection of soft materials such as weeds, moss, feathers, and cloth, usually placed in a tree. It eats mostly aerial insects. **Voice:** A loud chittering *kitterkitterkitter* and a short *kt'zee kt'zee*.

Breeding Info — **E**: 3–4 **I**: 14–16 **N**: 14–17 **B**: 1

Male

Female

Scissor-tailed Flycatcher
Tyrannus forficatus Length: 10"–15"
Wingspan: 15"

With the Scissor-tailed Flycatcher, it's all about
the tail — longest in the adult male, slightly
shorter in the adult female, and shortest (but
still remarkably long) in the juvenile. Also note
its pale gray head and underparts with colored
flanks — pinkish to orangish. When it flies it
reveals its salmon-colored underwings. It loves
to be in open country, where it perches on
exposed branches or posts to catch insects on
the ground or in the air. Nest is a loose collection
of natural and human-made materials placed on
a horizontal limb or on a telephone pole if no
trees are available. **Voice:** A short *pik* and a series
of short calls, like *wip wip wip wip*.

Juvenile

Breeding Info — **E:** 4–5 **I:** 12–15 **N:** 14–16 **B:** 1

Loggerhead Shrike

Lanius ludovicianus Length: 9" Wingspan: 12"

This shrike is common throughout the lower
48 states except in New England and the upper
Midwest. It is an efficient predator, capturing
its prey of frogs, lizards, insects, and small birds
and often impaling them on the thorns of trees
or the barbs on wire fences. Its hooked bill helps
it tear apart its prey. Look for its black mask,
wings, and tail contrasting with its gray back and
pale whitish underparts. During its undulating
flight, note the white patches on its outer wing
and white corners on its tail. Nest of twigs and
bark strips placed in a bush or tree. **Voice:** Harsh
sounds like *chhht* and rather musical notes like
chidlip tseedeedee julip julip.

Breeding Info—E: 4–7 **I**: 10–17 **N**: 17–21 **B**: 1–2

White-eyed Vireo

Vireo griseus Length: 5" Wingspan: 7½"

A beautiful vireo with yellow "spectacles" and a noticeable white iris. Often hidden in dense foliage, this species may best be found and recognized by its song, which is a varied warble with a harsh *check* before and after, much like "Quick! Give me the beer check!" Its flanks vary from pale greenish to yellow. Eats spiders, snails, lizards, and berries. Like many other vireos, it makes a small nest of woven plant fibers suspended by the rim from the fork of a horizontal branch. **Voice:** Besides song, it may give a nasal *jjheee* and a short *jwut*.

Breeding Info — **E:** 3–5 **I:** 12–16 **N:** 14 **B:** 1–2

Blue-headed Vireo

Vireo solitarius Length: 5½" Wingspan: 9½"

The vireos can be divided into two camps: those with "spectacles" and those without. Spectacles are a wide eye-ring connected to the bill base by a band of color. Blue-headed Vireos have white spectacles; White-eyed Vireos yellow spectacles; Red-eyed and Warbling Vireos have no spectacles. Spectacled vireos also have wingbars; those without do not. The Blue-headed has a bluish-gray head. It is mostly eastern, but has similar relatives in the West: the Cassin's and Plumbeous Vireos. Eats insects, spiders, and berries. Nest suspended from a horizontal branch. **Voice:** Song a series of short, well-spaced whistles like *teeyoo taree tawit*.

Breeding Info—E: 3–5 **I:** 11–12 **N:** 14 **B:** 1–2

Adult

Red-eyed Vireo

Vireo olivaceus Length: 6" Wingspan: 10"

The Red-eyed Vireo is a constant songster of deciduous forests in spring and summer. Listen for its repeated whistled phrases that we tend to tune out the next time you are walking or hiking in the woods. Unusual among most of our birds, the male Red-eyed continues to sing past the courtship and territorial phase and throughout breeding. Relatively unmarked, its best identifying features are its reddish eye, dark eyeline, and thin dark eyebrow (1st-winter birds have a dark eye). Gleans insects from tree leaves and eats berries. Cuplike nest of plant fibers is suspended by rim from horizontal branch. **Voice:** Well-spaced, whistled phrases like *teeawit teeyoo turaweet.*

1st winter

Breeding Info — **E**: 3–5 **I**: 11–14 **N**: 10–12 **B**: 1–2

Warbling Vireo

Vireo gilvus Length: 5½" Wingspan: 8½"

This is a very plain vireo of woods and shrubs
near water; inconspicuous if it were not for
its distinctive and continuous singing. Its song
is a long series of slurred whistles all strung
together, a little like a robin with a slide-whistle.
No wingbars, no spectacles, no red eye, and no
strong eyelines; basically you recognize it by
its lack of strong ID clues and its song. It is pale
beneath with variable yellow wash on the flanks.
Like most of the vireos, it has a nasal scolding
call like *neeyat*, which the bird often uses. Eats
insects and berries. Cuplike nest of bark strips
and grasses bound with spider webbing, suspended from fork on horizontal limb. **Voice:** As
described above.

Breeding Info – **E:** 3–5 **I:** 12–14 **N:** 12–16 **B:** 1–2

Steller's Jay

Cyanocitta stelleri Length: 11½" Wingspan: 19"

Our darkest jay, the Steller's is a denizen of coniferous forests in the coastal and mountainous
West. Its loud harsh calls may first alert you to its
presence in the dark woods. It has a black head
and back, deep blue wings and tail, and paler blue
belly; there are variable blue or white dashes on
forehead and over eyes. Often lives at higher
elevations in summer, then moves down to lower
elevations in winter, depending on availability
of food. Eats acorns, pine nuts, insects, berries,
nestlings, frogs, and comes to feeders. Nest of
twigs, leaves, mud lined with rootlets, placed in
tree or shrub. **Voice:** A rapid-fire *shushushushook*,
drawn-out *chjjj*, and grating *gr'r'r'r*.

Breeding Info – E: 2–6 **I:** 16 **N:** 18 **B:** 1

Blue Jay

Cyanocitta cristata   Length: 11"   Wingspan: 16"

A common and boisterous bird of woods and suburbs in the eastern two-thirds of the United States and southern Canada. Blue Jays are a gorgeous bird of deep blues and aquamarines that often go unappreciated. Some consider them bullies at feeders, but on the other hand, they are vigilant sentinels of the backyard, quick to warn other birds of possible danger. In fall Blue Jays store seeds and acorns under leaves on the ground, to be used as backup later in winter. Nest of twigs, bark, and leaves placed in a tree. **Voice:** Familiar *jay jay*, bell-like *tooloodle*, and grating *gr'r'r'r*.

Breeding Info — **E**: 4–5   **I**: 17   **N**: 17–19   **B**: 1–2

Western Scrub-Jay

Aphelocoma californica Length: 11½"
Wingspan: 15½"

By far the most common and widespread of our three North American scrub-jays. The other two, the Florida and Island Scrub-Jays, have very limited ranges in Florida and the Santa Cruz Islands off southern California. This is the backyard and feeder jay of most of the West and can become quite tame at picnic areas, parks, and feeders. Blue upperparts with a gray back, white eyeline, and white underparts, with a blue or dusky necklace. Eats a wide variety of foods including frogs, insects, eggs, acorns, berries; comes to feeders for seeds. Nest of twigs and grasses lined with rootlets placed in a shrub or tree. **Voice:** Includes scratchy *chee chee chee*, rising *jreet*, and percussive *gr'r'r'r*.

Breeding Info — **E**: 2–6 **I**: 16–19 **N**: 18 **B**: 1–2

Black-billed Magpie

Pica hudsonia Length: 19" Wingspan: 25"

A conspicuous and beautiful bird of the West,
the Black-billed Magpie is easily recognized by
its shape and colors. It has a thick crowlike bill,
very long graduated tail, and long, broad, pad-
dle-shaped wings; its head and back are black,
wings and tail primarily iridescent blue, and its
belly and scapulars white; white wing tips are
obvious in flight. Magpies eat a wide variety of
foods, from insects to berries to nestlings to
carrion; come to feeders for seeds. The nest is a
huge collection of twigs, up to 3 ft. in diameter,
with a side entrance and an interior cup of mud
or dung lined with softer materials. The nest is
placed near the top of a tree or in a shrub. **Voice:**
Call is a harsh *jeet* given singly or in a series.

Breeding Info – **E**: 2–9 **I**: 14–23 **N**: 10 **B**: 1

American Crow

Corvus brachyrhynchos Length: 17½" Wingspan: 39"

Ubiquitous and well known, the American Crow rarely goes unnoticed. It is usually found in small family-oriented groups during breeding and large to superlarge groups during winter. In fall and winter crows begin to fly together each evening to communal roosts. These rivers of birds can be seen at sundown flying into the hub where they spend the night, sometimes in hundreds of thousands. Roosts are in trees in wild or urban areas; their function is still a mystery. Crows eat a wide variety of items and may come to feeders for seed on the ground. They can be told from ravens by their smaller size and higher calls. Nest of twigs placed high in trees. **Voice:** Wide variety of short to drawn-out *caws*; also grating rattles.

Breeding Info—**E**: 4–5 **I**: 18 **N**: 28–35 **B**: 1–2

Fish Crow
Corvus ossifragus Length: 15" Wingspan: 36"

This is our smallest crow, and it is primarily
coastal, although it may also venture inland
along large rivers and to lakes. It looks almost
identical to the American Crow but has a distinctive voice. Rather than drawn-out *caw*s, it
gives short nasal calls like *unh unh* or *unh huh*. Fish
crows are usually found in small to large groups
and are our only crow to regularly rise up on
thermals of air. Nest is of sticks and twigs lined
with bark and grass, and placed in a tree. May
breed in small loose colonies. Diet includes a
wide variety of foods, including carrion. **Voice:** A
short *unh* or *unh unh* and a rattle call.

Breeding Info— **E:** 4–5 **I:** 18 **N:** 21–25 **B:** 1

Common Raven

Corvus corax Length: 24" Wingspan: 53"

The raven is a bird of mythological proportions, considered by many northwestern tribes as the creator of the world, the bringer of the sun, moon, stars, and fire, and the provider of fresh water. At the same time, other myths consider it a trickster. This great diversity is a tribute to the depth of its character. Common Ravens can seem remote in the wild cliffs where they breed or playful and inventive in their aerial maneuvers. They eat a wide variety of foods, including animal, vegetable, and carrion. Their nest is a mass of twigs, branches, and rootlets, placed on a cliff, tree, or building. **Voice:** A low resonant *groowk*, a drawn-out *kraaah*, and various rattling sounds.

Breeding Info — **E**: 3–8 **I**: 21 **N**: 35–42 **B**: 1

Female

Male

Purple Martin

Progne subis Length: 8" Wingspan: 18"

The Purple Martin is synonymous with bird-houses, because throughout the East the birds have become dependent upon human-made Purple Martin multicompartment nest boxes. In the West, Purple Martins nest singly and use old holes made by woodpeckers in trees or saguaro cacti. Purple Martins are immensely popular in the East, where dedicated Purple Martin "landlords" cater to the birds' nesting needs. In winter Purple Martins leave the United States and Canada. The male is all dark with a glossy bluish-black body; the female has an all-light belly; 1st-year males have a blotchy belly. They eat aerial insects. Nest of mud, sticks, leaves (some green). **Voice:** A gurgling set of paired notes and calls like _chur_ or _chur chur_.

Martin house

Breeding Info — **E**: 5–6 **I**: 15–16 **N**: 27–35 **B**: 1

Male

Violet-green Swallow

Tachycineta thalassina Length: 5¼"
Wingspan: 13½"

The Violet-green is a small swallow of the West
named for its green back and the dark violet
rump of the male (the female has a sooty rump).
Both sexes are dark above and white below
with narrow white rectangular hip patches
that extend well onto the rump. The male has
a whitish face, the female a dusky face. When
they are perched, you can see the extremely
long wings that project well past the tail tip.
Violet-green Swallows eat aerial insects. The nest
of grasses lined with feathers is placed in a cavity
such as an old woodpecker nest, birdhouse, or
rock crevice. **Voice:** A harsh *choit*, a repeated
cheecheechee, and a high-pitched *tseet*.

Female

Breeding Info – **E**: 4–6 **I**: 13–14 **N**: 21 **B**: 1–2

Tree Swallow

Tachycineta bicolor Length: 5¾" Wingspan: 14½"

The Tree Swallow is well named for its habit of
nesting in tree holes, such as abandoned nests of
woodpeckers. They also can easily be attracted
to birdhouses placed in open areas, especially
near water. The adults are dark iridescent
bluish-green above and white below. 1st-year
females are mostly brown on the back. In fall
Tree Swallows gather into flocks of hundreds
to thousands, feeding on aerial insects and bay-
berries as they travel south. Most Tree Swallows
leave the United States in winter, but some
remain on the south Atlantic and Gulf coasts.
Nest is composed of grasses and feathers. **Voice:**
Includes a *chirdup chirdup*, a *cheedeep*, and *chee
chee chee* in a long series.

Breeding Info — **E**: 5–6 **I**: 14–15 **N**: 21 **B**: 1–2

Barn Swallow

Hirundo rustica Length: 6¾" Wingspan: 15"

The Barn Swallow is perhaps the most beautiful flier among our swallows. Its long pointed wings and long streamered tail seem to help it gracefully speed and arc through the sky. It is a handsome swallow, with the bluish-black of its upperparts and the pale to deep orange of its forehead and chin. Male and female build the nest of mud and grasses, usually in a barn or outbuilding, under a bridge or in a culvert. Nests may be used in successive years. **Voice:** A short *chevit* or *chevitvit* and a continuous twitter with grating sounds.

Breeding Info — **E:** 2–7 **I:** 14–16 **N:** 18–23 **B:** 1

Horned Lark
Eremophila alpestris Length: 7¼" Wingspan: 12"

Horned Larks are attracted to open areas with
sparse vegetation, whether a field, prairie, air-
port, or shoreline. The "horns" are nothing more
than feather tufts that project off the back of the
crown, generally larger on males. Horned Larks
have a unique facial pattern of black and yellow,
and a dark bib across the breast. They walk along
the ground as they feed on insects and seeds,
often briefly flying up in a flock. Nest is a scrape
lined with grasses, often placed near a tuft of
grass. **Voice:** Song is a rising trill; common flight
call is *tsee titi*.

Breeding Info — **E**: 3–5 **I**: 11 **N**: 9–12 **B**: 1–3

Male, coastal subspecies

Bushtit

Psaltriparus minimus Length: 4½" Wingspan: 6"

The Bushtit is a very small, round-bellied, long-tailed bird with a short stubby bill. It has two main forms: those with a brown crown (coastal subspecies) and those with a gray crown (interior subspecies). Males have dark eyes and females have pale eyes. They are acrobatic feeders, hanging upside down as they glean insects from leaves. They also eat seeds and come to bird feeders. In winter they move about in small flocks of 6 to 30 or more birds, keeping in contact with continuous chatter. Their gourd-shaped nest of mosses and rootlets is suspended from the twigs of a shrub or tree. **Voice:** A loud high-pitched *pseeet*, a clipped *chitchit*, and a short *tsit*.

Female, interior subspecies

Breeding Info—E: 5–7 **I**: 12 **N**: 14–15 **B**: 1–2

Carolina Chickadee

Poecile carolinensis Length: 4¾" Wingspan: 7½"

The Carolina Chickadee is the alter ego of the
Black-capped Chickadee. They look essentially
the same and have similar habitats and behavior.
They are best told apart by their songs and calls.
The Carolina's song is a 4-part whistle like *feebee
feebay*, while the song of the Black-capped is
just a 2-part *feebee*. The calls of the Carolina are
generally higher, harsher, and given more rapidly
than those of the Black-capped. They can excavate their own nest holes in soft wood and may
nest in birdhouses. In the wild, they eat insects
and seeds; come to feeders for sunflower seed
and suet. **Voice:** Song as described; calls include
tseedleedeet, *chickadeedeedee*, and *tseet*.

Breeding Info — **E**: 6 **I**: 11–12 **N**: 13–17 **B**: 1–2

Black-capped Chickadee

Poecile atricapillus Length: 5¼" Wingspan: 8"

Black-capped Chickadees are the regulars of any
bird-feeding station. They are always the first
to visit any new feeder and the first to exercise
their curiosity on any changes you might have
made. Chickadees are found in pairs during
breeding and in small flocks the rest of the time.
During spring you can hear males countersing
their whistled *feebee* songs as they try to outline
territories and attract mates. They excavate
their nest holes in trees with softer wood, such
as those softened by fungi; they also nest in
birdhouses. Eat insects and seed; eat sunflower
seeds and suet at feeders. **Voice:** Song is a 2-part
whistled *feebee*; also give *tseedleedeet, chebeche,
chickadeedeedee,* and *tseeet.*

Breeding Info—E: 6–8 **I:** 12 **N:** 16 **B:** 1–2

Chestnut-backed Chickadee

Poecile rufescens Length: 4¾" Wingspan: 7½"

The Chestnut-backed Chickadee is the chickadee of the West Coast and the northern Rockies. It is easily recognized by its brown cap and chestnut back and flanks. Chestnut-backeds form small flocks in winter that roam about a fixed area. They are often joined temporarily by other species such as kinglets, nuthatches, and titmice, possibly so they have more eyes looking out for danger. They use a tree cavity, excavated hole, or birdhouse for their nest, which is made of hair, moss, and other plant fibers. They eat insects and seeds and come to feeders for sunflower seeds and suet. **Voice:** They seem to lack the whistled song of the other chickadees. Calls include a buzzy *chickazeezeezee, seebeeseebee,* and *dzeeet.*

Breeding Info— **E**: 6–7 **I**: 11–12 **N**: 13–17 **B**: 1–2

Mountain Chickadee

Poecile gambeli Length: 5¼" Wingspan: 8½"

This is the chickadee of the western interior mountains. Its favorite habitat is open coniferous woods; in winter it may move to slightly lower elevations. It is the same size as the Black-capped Chickadee but has a distinctive white eyebrow. Its whistled song is also different; it is 2 to 6 whistled notes on either the same or different pitches, like *feefeefee* or *fee bay bay*. It feeds on insects and seeds, and comes to feeders for sun-flower seed and suet. Nest of wood chips, hair, and feathers in an excavated or natural cavity or birdhouse. **Voice:** Calls similar to Black-capped's but higher-pitched and faster.

Breeding Info — **E**: 7–9 **I**: 14 **N**: 17–20 **B**: 1–2

Adult

Tufted Titmouse

Baeolophus bicolor Length: 6½" Wingspan: 9¾"

The Tufted Titmouse is the only titmouse in the East. It is a little gray bird with a relatively long tail and prominent crest; it has a black patch on its forehead and variable orange-buff flanks. Its habits are similar to those of chickadees — found in pairs during breeding and in small flocks the rest of the year. Titmice love feeders, and after chickadees will be the next visitor to any new feeder. The song of the male rings out through the moist air of spring mornings; it is a 2-note downslurred whistle, like *peter peter peter*. Titmice eat insects, seeds, suet, and berries, and nest in tree cavities. **Voice:** Song as above. Calls include an insistent *jwee jwee jwee*, a *see see see*, and a soft *tseeet*.

Juvenile

Breeding Info — E: 4–8 **I:** 13–14 **N:** 17–18 **B:** 1–2

Oak Titmouse

Baeolophus inornatus Length: 5¾" Wingspan: 9"

The Oak Titmouse lives in California and southern Oregon, and is a favorite bird of gardens and feeders. It is a very plain bird — gray all over with a large dark eye — and has an obvious crest. It is similar to its close relative the Juniper Titmouse, which lives in mixed woods with junipers but has a smaller bill, browner back, more western range, and different song. Eats seeds, nuts, and insects, and comes to feeders for seed and suet. Nests in tree cavities, woodpecker holes, and birdhouses. **Voice:** Song is a series of whistled 2–note phrases like *peyer peyer* (Juniper song is a rapid trill); calls include *tsikajwee, see see see,* and a soft *tseet.*

Breeding Info — **E:** 6–8 **I:** 14–16 **N:** 16–21 **B:** 1–2

Male

White-breasted Nuthatch

Sitta carolinensis Length: 5¾" Wingspan: 11"

The White-breasted Nuthatch is our most widespread and common bird that walks headfirst down tree trunks. This is how it looks for insects in tree bark as it forages. When not foraging on tree trunks and limbs, they are often storing food in bark crevices for later use. They are small with a fairly long beak; white beneath and gray above with a narrow dark cap (black on the male, more silver on the female); they have a variable touch of rust on their rear flanks. The nest of downy materials is placed in a natural cavity, old woodpecker hole, or birdhouse. **Voice:** Eastern birds give a rapid *werwerwer* and well-spaced *ank ank*. Western birds give higher-pitched and raspier calls.

Female

Breeding Info — **E**: 3–10 **I**: 12 **N**: 14 **B**: 1–2

Male

Red-breasted Nuthatch

Sitta canadensis Length: 4½" Wingspan: 8½"

Red-breasted Nuthatches are more attached to evergreen forests of the North and West and higher elevations than their larger cousin the White-breasted Nuthatch. They are a colorful bird: rusty to orangish-red below, smooth gray above, a white face with a thin black eyeline, and narrow dark cap (black in the male, bluish-gray in the female). They also can walk headfirst down tree trunks, and they eat conifer seeds and insects and come to feeders for sunflower seed and suet. Their nest of mosses, shredded bark, and grasses is placed in an excavated hole, abandoned woodpecker hole, or birdhouse. **Voice:** Song is a nasal *meep meep meep*; calls include a soft *ink ink*.

Female

Breeding Info—E: 5–7 **I:** 12 **N:** 16–21 **B:** 1–2

Brown Creeper

Certhia americana Length: 5¼" Wingspan: 7¾"

This can be a hard bird to find, for its streaked
brown back can make it camouflaged against
the tree bark on which it makes its home. Brown
Creepers, true to their name, hitch up tree
trunks in a spiral fashion, and when near the top,
drop down to the base of the next tree and start
all over again. They hold on to the bark with
their feet and rest on their tail feathers, just like
woodpeckers. Their thin downcurved bills help
them probe for small insects in bark crevices.
They will come to feeders for suet. Their nest of
mosses and bark strips is often placed behind a
large loose piece of bark. **Voice:** Calls include a
very high-pitched, quavering *tseee* or *tseetseeee*.

Breeding Info — **E**: 5–6 **I**: 14–16 **N**: 13–15 **B**: 1

Cactus Wren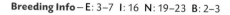

Campylorhynchus brunneicapillus
Length: 8½" Wingspan: 11"

This is a large and expressive wren of southwestern deserts. It is not afraid to perch out in the open atop a cactus and give its rather unmusical song. It is recognized by its large size, reddish-brown crown, white eyebrow, and spotted breast. Cactus Wrens feed on the ground, eating insects, spiders, lizards, seeds, and berries; may come to feeders for seed and fruit. Large globular nest of plant stems and grasses has a side entrance leading to an inner chamber lined with fur and feathers; nest is placed in cholla or other prickly plant. **Voice:** Song a long series of harsh *chut*s or *churit*s; calls include a rattling chatter and quacking *krak krak*.

Breeding Info—E: 3–7 **I:** 16 **N:** 19–23 **B:** 2–3

House Wren
Troglodytes aedon Length: 4¾" Wingspan: 6"

With its cheery bubbling song and attraction to
backyard birdhouses, who would not welcome
the presence of this active little bird? What
House Wrens lack in color, they make up for in
personality. Upon arriving in spring, the male is
a busy builder, stuffing short twigs into numerous nest holes. When the female shows up, she
chooses one of the nests and adds her own fin-
ishing touches. The House Wren is plain-faced;
brown above and paler below, with barring on
wings and tail. It eats insects and spiders found
on the ground or gleaned from leaves. **Voice:**
Song is a cascading jumble of fairly harsh trills;
calls include a chattering *ch'ch'ch'ch*.

Breeding Info — **E**: 5–6 **I**: 12–15 **N**: 16–17 **B**: 1–2

Carolina Wren

Thryothorus ludovicianus Length: 5½"
Wingspan: 7½"

The bright melodic song of the Carolina Wren
can be heard all year long. Repeated whistled
phrases like *teakettle teakettle teakettle* ring out
from backyard to backyard where this wren
makes its home. It is one of our most colorful
wrens, rich buff below, warm brown above, with
a bright white eyebrow and throat surrounding
a dark eyeline. It forages on the ground and in
shrubs, looking for insects, spiders, and snails. It
is quirky in its choice of nesting spots, accepting
any nook in a building, tree cavity, or human
object left outside, like a flowerpot, old boot, or
basket. **Voice:** Calls include a chattering *churrrr*
and a 2-part *tidik*.

Breeding Info — **E**: 4–8 **I**: 12–14 **N**: 12–14 **B**: 1–3

Ruby-crowned Kinglet

Regulus calendula Length: 4¼" Wingspan: 7½"

The Ruby-crowned Kinglet is a very small bird
that rarely sits still. It is always making short
flits from branch to branch, constantly flicking
its wings as it searches for insects on leaves and
limbs. It is grayish-olive above and paler below;
its large dark eye has white crescents in front
and behind, and there is a distinct dark area
below the lower wingbar; the ruby crown is
the male's and is usually concealed. Its nest of
mosses and lichens is suspended from a branch.
Voice: Song a high-pitched whistled *see see see
look at me, look at me, see see see.*

Breeding Info — **E**: 5–11 **I**: 12 **N**: 12 **B**: 1

Golden-crowned Kinglet
Regulus satrapa Length: 4" Wingspan: 7"

The Golden-crowned Kinglet is even smaller than the Ruby-crowned but much easier to identify, because its golden crown is always visible and is flanked by black crown stripes that join on the forehead. Like the Ruby-crowned, it is always active and constantly flicks its wings as it forages for insects and spiders in shrubs and trees. It breeds in coniferous woods and winters in deciduous or mixed woods. Its nest is a globular collection of mosses, lichens, and spider silk lined with feathers and suspended from a branch.
Voice: Song is a series of high-pitched *tseees*, often rising; calls include a high-pitched *dzeee*.

Breeding Info – E: 8–11 I: 14–15 N: 14–19
B: 1–2

Blue-gray Gnatcatcher
Polioptila caerulea Length: 4½" Wingspan: 6"

Of the three gnatcatchers in the United States
and Canada, this is the most common and widespread. It breeds in most of the lower 48 states
and winters in the southern United States. It is
gray to bluish above, whitish below, with a thin,
complete white eye-ring. It is long billed and
long tailed for such a small bird and constantly
waves its tail from side to side. It catches insects
by gleaning, sallying out to catch them in the air,
and by briefly hovering in front of branch tips
to pick them off outer leaves. Nest a tiny cup of
plant fibers and spider silk coated with lichens.
Voice: Common call a nasal *tzyeee*; song a long
high-pitched twittering, sounding a little like a
muted catbird.

Breeding Info — **E**: 3–6 **I**: 13 **N**: 10–12 **B**: 1–2

Male

Female

Eastern Bluebird

Sialia sialis Length: 7" Wingspan: 13"

The Eastern Bluebird is a much beloved bird. The male's upperparts are brilliant blue, prompting Thoreau to say, "The bluebird carries the sky on his back." In addition he is red and white below—what patriotic colors! Eastern Bluebirds' acceptance of birdhouses for breeding and a movement to put up thousands of bluebird houses across the country have saved this species from a serious decline in the late 1900s. The male has a brilliant blue back; the female is grayish blue on the back; the young have spots on the breast. Come to feeders for suet, sunflower seed, and mealworms. Song likened to *cheer cheerful charmer*; calls include a whistled *turwee* and a chattered *ch'ch'ch'ch*.

Juvenile and adult female

Breeding Info—**E:** 3–6 **I:** 12–18 **N:** 16–21 **B:** 1–3

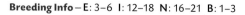

Male

Western Bluebird

Sialia mexicana Length: 7" Wingspan: 13½"

The Western Bluebird population was seriously declining in the mid 1900s but started to turn around in the 1970s, when dedicated individuals and small hardworking societies made concerted efforts to put up and monitor thousands of birdhouses. The Western Bluebird breeds from the Northwest through California to the Southwest. The male has a blue head and throat (Eastern male has a red throat) and chestnut on the sides of his back. The female head and throat are all grayish, with no strong contrasts. Nest of grasses, pine needles, and sometimes feathers placed in a cavity or birdhouse. **Voice:** Song a *ch'ch'ch churrr chup*; calls include *chweer* and harsh *ch'ch'ch*.

Female

Breeding Info—E: 5–8 **I:** 13–17 **N:** 19–22 **B:** 1–3

Male

Mountain Bluebird

Sialia currucoides Length: 7¼" Wingspan: 14"

The Mountain Bluebird prefers more open land than do its relatives, living in prairies, sagebrush, grasslands, and barren fields, where it often hovers above the ground while searching for insects to capture. The male is a striking turquoise overall except for a silvery belly and undertail. The female is the least colorful of the bluebirds but has turquoise wings. In the open habitats it favors there are fewer tree cavities for nesting, so Mountain Bluebirds may use rock crevices, nooks in buildings, old Cliff Swallow nests, and birdhouses. Nest is made of grasses, bark shreds, and pine needles, lined with hair or feathers.
Voice: Calls and song contain low-pitched rough sounds like *churrr*.

Female

Breeding Info – **E**: 5–8 **I**: 13–17 **N**: 19–22 **B**: 1–3

Hermit Thrush
Catharus guttatus Length: 6¾" Wingspan: 11½"

All of the thrushes are skilled songsters that are more often heard than seen. The Hermit Thrush's song is a lovely flutelike series of rising and falling notes with a distinctively extended first note. This is our one thrush that typically overwinters within the United States. It has the helpful identifying trait of flicking its tail quickly up and slowly down. Also, its tail is the reddest portion of its upperparts. Feeds on ground, eating insects, spiders, worms, and snails. Nest of mosses and grasses lined with rootlets and bark strips is placed from on the ground to 8 ft. high. **Voice:** Calls include a rising *zweee* and a short *chup*.

Breeding Info — **E**: 3–6 **I**: 12–13 **N**: 12 **B**: 1–2

Wood Thrush

Hylocichla mustelina Length: 7¾" Wingspan: 13"

The gorgeous song of the Wood Thrush is a
lovely flutelike series of rising and falling notes
with an introductory stutter, like *but but but
eeyolay...but but but aholee*. It is our largest thrush,
recognized by its black-and-white streaked face
and its reddish head and back that blend to
duller reddish-brown on rump and tail. This is
a bird of mixed deciduous woods with a shrub
understory. It feeds mostly on the ground, eating
earthworms, insects, snails, and fruits. Its nest
of mosses, leaves, and string has a middle layer
of mud, then a lining of rootlets, and is placed
on a forked branch 5–20 ft. high. **Voice:** Calls
include a murmuring *bwubwubwub* and a louder
bweebweebweeb.

Breeding Info – E: 2–5 **I:** 12–13 **N:** 12–14 **B:** 1–2

Swainson's Thrush
Catharus ustulatus Length: 7" Wingspan: 12"

The song of the Swainson's Thrush is a lovely ascending spiraling whistle like *yooray youree youreeyee*. The bird can be recognized by its uniformly grayish-brown upperparts and its buffy eye-ring connected to its bill with a buffy bar, creating a spectacled look. It also has a faintly buffy upper breast and buffy arc under the side of the face. Swainson's like coniferous and deciduous woods in the East, streamside woods in the West. Feeds mostly on the ground, eating spiders, insects, and fruits. Nest of twigs, mosses, and grasses is placed on a horizontal limb 2–20 ft. high. **Voice:** Calls include a flat *kweeet*, a short *kwip*, a 2-part *kw'purr*, and a high whistled *peeet*.

Breeding Info — **E**: 3–5 **I**: 11–14 **N**: 10–14 **B**: 1

Veery

Catharus fuscescens Length: 7" Wingspan: 12"

The Veery's beautiful song is a descending, spiraling, flutelike whistle like *tooreeyur-reeyur-reeyur-reeyur*. The similar Swainson's song ascends. The Veery is the cinnamon thrush; its upperparts are uniformly light cinnamon (darker and duller in the West), as are the small spots limited to its buffy upper breast. Veerys feed on the ground in moist deciduous woods, often near streams, and eat insects, spiders, snails, and fruit. Nest of weed stems and mosses lined with dead leaves and rootlets placed on or near the ground. **Voice:** Calls include a descending *veeer* and *veerit*.

Breeding Info — **E**: 3–5 **I**: 10–12 **N**: 10 **B**: 1–2

Adult

Adult

American Robin

Turdus migratorius Length: 10" Wingspan: 17"

We tend to think of robins as just backyard birds, but they actually live in a wide variety of habitats, from urban to wild and valley to mountaintop. Almost everybody can identify a robin by its brick red underparts and charcoal upperparts. The young are similar but with dark spots on the breast. They typically take short runs punctuated by brief stops as they look for earthworms and insects to eat. In fall and winter they often gather into large flocks, roosting together at night and dispersing to eat (mostly berries) during the day. Nest of mud and grasses placed in tree or on building ledge. **Voice:** Song is a whistled *cheeryup cheerily*; calls include a low *tuk tuk tuk* and a high *teeek*.

Nestlings

Breeding Info — **E**: 3–7 **I**: 12–14 **N**: 14–16 **B**: 1–3

Brown Thrasher

Toxostoma rufum Length: 11" Wingspan: 13"

The Brown Thrasher is both boldness and
stealth: the male loudly singing double-phrased
songs from a conspicuous perch, or the pair
skulking around dense shrubbery, giving us just
a glimpse. They are easily recognized by their
long tails, downcurved bills, and uniformly
reddish-brown upperparts. They feed on the
ground, tossing leaves aside in search of insects,
lizards, frogs, and berries. Nest of twigs, leaves,
and grapevines lined with rootlets is placed on
ground or up to several feet high. **Voice:** Song
a long series of whistled phrases, each usually
repeated twice and often containing imitations
of other birds. Calls include a kisslike *smack* and a
harsh rising *peeyee*.

Breeding Info — **E:** 2–6 **I:** 12–14 **N:** 9–12 **B:** 1–2

Curve-billed Thrasher

Toxostoma curvirostre Length: 11"
Wingspan: 13½"

The Curve-billed is the familiar thrasher of
the Southwest, where its loud *wit-weeet* call is
a common sound in desert areas. Among the
similar-looking thrashers, the Curve-billed has
a medium-length moderately downcurved bill
that is usually all dark. The California, Crissal,
and Le Conte's Thrashers all have longer, more
strongly downcurved bills. Curve-billeds eat
insects on the ground and come to bird feeders
for fruit. Nest of twigs, grasses, and leaves lined
with horsehair and rootlets is placed in a thorny
bush or cactus. **Voice:** Song is a mixture of whistles and sounds, with some imitation. Calls also
include a chattering *churrr*.

Breeding Info – E: 1–5 **I:** 12–15 **N:** 12–18 **B:** 1–3

Gray Catbird

Dumetella carolinensis Length: 8½" Wingspan: 11"

The Gray Catbird is named for its meowlike
call. It is easily recognized, a medium-sized gray
bird with a black cap. Its only brighter color is a
reddish-brown undertail seen as the bird cocks
its tail. The catbird, mockingbird, and thrashers
are related and are known for their songs, which
contain imitations of other birds' calls and songs.
In general mockingbirds repeat phrases 3 times,
thrashers twice, and catbirds once, so count
the imitations to identify the bird! Catbirds eat
insects and berries; come to feeders for fruit and
mealworms. The nest of twigs, leaves, and grape
bark lined with rootlets is placed in shrubbery.
Voice: In addition to song and call mentioned
above, a soft *kwut* and a ratchety *ch'ch'ch'ch*.

Breeding Info – E: 2–6 **I:** 12–14 **N:** 10–13 **B:** 1–2

Northern Mockingbird

Mimus polyglottos Length: 10½" Wingspan: 14"

The Northern Mockingbird is best known for its
song, a long series of imitations of other birds'
sounds. It tends to repeat each imitation 3 or
more times. It also may imitate other sounds,
like those of frogs, fire engines, dogs, and many
more. Male mockingbirds sing in spring to
attract a mate and defend a nesting territory;
both sexes sing in fall when defending a winter
feeding territory. Mockingbirds may also sing on
a moonlit night or if near a streetlamp. They eat
insects during breeding and berries the rest of
the year. Nest of twigs, grasses, and bark lined
with rootlets is placed in shrubbery. **Voice:** In
addition to song, calls include a deep *chewk*, a
rapid *ch'ch'chik*, and a harsh *chjjjj*.

Breeding Info – **E:** 2–6 **I:** 12–13 **N:** 10–13 **B:** 1–3

Cedar Waxwing

Bombycilla cedrorum Length: 7" Wingspan: 12"

Cedar Waxwings are almost always seen in flocks. They eat insects and berries in summer and berries almost exclusively in winter. Their thin, very high-pitched *dzeee* call is a good way to recognize their presence, but it can easily go unnoticed by a novice birder. They are brownish-gray above with a yellow belly and yellow-tipped tail. Their feathers blend perfectly, making it hard to see any individual feather and giving the bird an unblemished sleekness. They have a short thin crest, and the bright red waxy tips on the inner wing give them their name. Nest of grasses, twigs, and moss lined with rootlets is placed 4–50 ft. high. **Voice:** Call mentioned above.

Breeding Info—**E:** 2–6 **I:** 12–16 **N:** 14–18 **B:** 1–2

Adult summer

European Starling
Sturnus vulgaris Length: 8½" Wingspan: 16"

The European Starling is the consummate city
bird, able to feed on city streets, nest in nooks of
buildings or stoplights, and roost under bridges
in winter. Introduced into New York City from
Europe in the 1890s, it now thrives throughout
most of North America. Starlings in summer are
glossy black with some light spotting, and have a
yellow bill and red legs; in winter they are heavily
spotted and have a dark bill and dark legs into
December. Winter is their time to gather into
large flocks that feed and roost together. Starlings eat insects, spiders, earthworms, berries,
and food scraps. They nest in nooks and crannies. **Voice:** Song is a rambling mix of squeals
and whistles, often with calls of other birds. Calls
include a downslurred whistle like *wheeeuuu*.

Adult winter

Breeding Info — **E**: 2–8 **I**: 12–14 **N**: 18–21 **B**: 1–3

Ovenbird

Seiurus aurocapilla Length: 5¾" Wingspan: 9½"

The Ovenbird is one of our ground-dwelling warblers, spending most of its time walking on the forest floor, flipping over leaves to look for insects, worms, and spiders. It also builds its nest on the ground, a domed-over construction of grasses and rootlets resembling a Dutch oven, thus the bird's common name. Ovenbirds are olive-brown above and white with broad blackish streaking beneath. On the head, two black crown stripes border a central orange streak. Their legs and feet are pinkish. **Voice:** Song is a loud repeated phrase like *teacher teacher teacher;* call is a sharp *chit.*

Breeding Info — **E**: 3–6 **I**: 11–14 **N**: 8–11 **B**: 1–2

Male

Black-and-white Warbler

Mniotilta varia Length: 5¼" Wingspan: 8¼"

This is a handsome warbler that can hitch about on tree trunks and limbs in search of insects. It is the zebra warbler, with black and white streaking all over. The male has a black throat and cheek; the female a white throat and buffy cheek. Its relatively long bill helps it pry insects out of deep bark crevices. It is another warbler that overwinters along the Southeast and Gulf coasts. Nest a grassy cup lined with leaves and moss placed on the ground under cover of a tree. **Voice:** Song a long series of high repeated 2-note phrases like *weesy weesy weesy*; call a sharp *pit*.

Breeding Info — **E**: 4–5 **I**: 10–13 **N**: 8–12 **B**: 1–2

Female

Male

Common Yellowthroat

Geothlypis trichas Length: 5" Wingspan: 6¾"

This is the most widespread breeding warbler
in North America, favoring moist roadside
ditches, brushy edges, and marshy shorelines.
The male looks like a little bandit with his black
mask, and his song sounds as if he's holding you
up – "Your money, your money, your money."
Both sexes have brownish-olive upperparts and
yellow underparts, brightest on the throat. The
female lacks the mask. This is an active warbler,
constantly hopping about between branches
and flicking and cocking its tail. Common Yel-
lowthroats eat insects off leaves. Nest of downy
material placed in vertical fork of small shrub or
tree 4–15 ft. high. **Voice:** Calls include a distinctive *tchat*.

Female

Breeding Info – E: 3–4 **I:** 12 **N:** 8–9 **B:** 1–2

Male

Northern Parula

Setophaga americana Length: 4¼" Wingspan: 7"

The Northern Parula is small even by warbler standards, with a noticeably short tail. Its head and upperparts are bluish-gray with two short white wingbars and an olive upper back. Beneath, it is white with a yellow breast and yellow lower bill. The male has variable black and chestnut bands across his yellow chest. Their favorite habitat is forests with Spanish moss (in the South) and with usnea lichens (in the North), for these materials are used in their nests. They eat insects gleaned off leaves. **Voice:** Song a buzzy ascending trill that ends with an abruptly lower note, like *zeeeee-yup*; call is a musical *djip*.

Breeding Info — **E**: 3–7 **I**: 12–14 **N**: 10–11 **B**: 1

Female

Male

Yellow Warbler

Setophaga petechia Length: 4¾" Wingspan: 8"

The operative word for this warbler is "yellow," for no other North American warbler is so completely yellow. The male has obvious reddish streaking on his underparts, the female more subtly so. This species loves moist shrubby areas, usually near streamsides or in swamps, and shuns the deep forests favored by most of its relatives. The nest is a tight cup of milkweed stem fibers, willow down, and fine grasses placed in the upright fork of a shrub, especially a willow. This is one of the first warblers to start migrating south, often beginning in late July. **Voice:** A musical series of high whistles accented at the end, like "Sweet sweet sweet I'm so sweet"; call is a fairly loud *tchip*.

Female

Breeding Info — E: 4–6 **I:** 10 **N:** 9–11 **B:** 1

Male

Magnolia Warbler

Setophaga magnolia Length: 4¾" Wingspan: 7½"

The Magnolia Warbler in summer seems to have collected ID clues from all the other warblers and put them on itself. It has a black mask, white eyebrow, yellow belly, black streaking suggesting a necklace across its chest, and a unique undertail pattern of white tail with a black tip. The female is similar but less heavily marked. In winter Magnolias change plumage as the mask and white eyebrow are replaced by a grayish head and thin white eye-ring, and the streaking on their underparts becomes only faintly visible. They feed on insects and spiders gleaned from leaves. Nest of twigs and grasses placed in a conifer. **Voice:** Song is a variable short series of musical notes like *weeti weeti weetio*, and call is a nasal *clenk*.

Female

Breeding Info – E: 3–5 **I**: 11–13 **N**: 8–10 **B**: 1

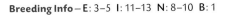

Summer male (Myrtle)

Winter (Myrtle)

Yellow-rumped Warbler

Setophaga coronata Length: 5¼" Wingspan: 9¼"

The most common warbler seen during migration, especially in fall, recognized in all plumages
by its bright yellow rump ("butter butt") and
yellow patches flanking its breast. Winter birds
are duller than summer birds, females duller
than males, and 1st-winter females the plainest
of all, sometimes with no yellow on the breast.
The Yellow-rump subspecies in the West is
sometimes referred to as Audubon's Warbler
and usually has a yellow throat; the eastern subspecies is sometimes called Myrtle Warbler and
has a white throat. Eating insects in summer and
berries in winter, they are our most widespread
overwintering warbler. Nest of twigs and grasses
placed in a conifer. **Voice:** Song is a weak variable
trill; calls include a sharp loud *check*.

Summer male (Audubon's)

Breeding Info – **E**: 4–5 **I**: 12–13 **N**: 12–14 **B**: 1–2

Male

Black-throated Green Warbler

Setophaga virens Length: 4¾" Wingspan: 7¾"

This is a colorful warbler with a memorable song sounding much like *zee zee zee zoo zee*, with the last note highest in pitch. The summer male has a solid black chin, throat, and upper breast contrasting with a bright yellow face. The summer female is similar but with a whitish to yellowish chin. Winter birds have less black on their underparts. Black-throated Greens forage at medium to high levels in trees, searching for insects and berries. The nest of grass and bark shreds is lined with feathers and hair and is placed high in a conifer. **Voice:** Call is a loud *tsik*.

Breeding Info — **E**: 4–5 **I**: 12 **N**: 8–10 **B**: 1

Female

Summer, male

Townsend's Warbler
Setophaga townsendi Length: 4¾" Wingspan: 8"

This popular western warbler breeds in the Pacific Northwest, migrates throughout the West, and can be found wintering on the coast from Washington through California. The male has the most contrasting plumage, with a black cap, ear patch, and throat highlighting its yellow face; the female is similar but the black areas are a duller slate. In winter both sexes are duller. Townsend's eat insects and spiders and occasionally visit feeders for suet, nuts, and fruit. Nest of grasses and twigs is placed at the top of tall conifer. **Voice:** Song a series of whistled and buzzy notes ending on a higher pitch; calls include a sharp *tip*.

Breeding Info—**E:** 3–5 **I:** 12 **N:** 8–10 **B:** 1

Winter

Male

Black-throated Blue Warbler

Setophaga caerulescens Length: 5" Wingspan: 7¾"

This is the warbler with a white handkerchief in its pocket. It is really just white at the base of the outer wing feathers, but when the bird is perched, it looks like a pocket handkerchief. Males are a striking bluish above with a black face, throat, and flanks and a white central belly. The female is a plainer brownish-olive above and buffy to yellow below, but just look for her handkerchief. The Black-throated Blue is a bird of the understory, where it feeds on insects, seeds, and fruit. Bulky nest of bark strips and wood fibers placed low within a conifer or shrub. **Voice:** Song is 3 to 5 buzzy notes with the last upslurred, like *zoo zoo zoo zeee*; call is a harsh *kik*.

Breeding Info — **E**: 3–5 **I**: 12 **N**: 10 **B**: 1

Female

Male

Prairie Warbler
Setophaga discolor Length: 4¾" Wingspan: 7"

The Prairie Warbler is mainly a southeastern warbler that can be found overwintering in Florida and along the Gulf Coast. Its call is a series of rising buzzy notes, sounding like a boiling teakettle. Overall, the Prairie is a yellow warbler with some strong black streaking on the sides of the breast and flanks, and a little yellow arc under its eye. Prairies in their 1st year have a whitish arc under the eye. The sexes are similar with female slightly paler. The Prairie has the habit of bobbing its tail, an additional helpful ID clue. They eat insects gleaned from leaves or caught in the air. Nest of grasses and plant down placed in bushes or small trees. **Voice:** Calls include a musical *tsip*.

Female

Breeding Info — **E**: 3–5 **I**: 10–14 **N**: 8–10 **B**: 1–2

Male

Pine Warbler

Setophaga pinus Length: 5¾" Wingspan: 8¾"

The Pine Warbler is almost always in or near pines. In these areas you can hear its musical trill drifting down from the tops of the trees. The male is mostly bright yellow with a white under-tail, white wingbars, and a little streaking on the breast; it also has spectacles created by a yellow eye-ring with a connecting bar to its bill. The female is similar but with duller and less extensive yellow. 1st-year females can look almost all gray. Pine Warblers hitch around tree trunks and branches in search of insects; they may also visit feeders for seed and suet. Nest of pine needles and weed stems placed high on horizontal pine branch. **Voice:** Call a slurred rich *chip*.

Female

Breeding Info—**E**: 3–5 **I**: 10 **N**: 10 **B**: 1

Summer (Eastern Palm)

Palm Warbler

Setophaga palmarum Length: 5½" Wingspan: 8"

The Palm is a common warbler overwintering in the Southeast but is also seen on migration to Canada, where it breeds. It has an olive-yellow rump and bright yellow undertail, which makes it look like it sat in mustard and is trying to flick it off with its characteristic tail-bobbing. In summer Palm Warblers additionally have a rusty cap and yellow throat. In winter they are duller, lack the rusty crown, and have variable yellow on the chin, but note their tail-bobbing and "mustard." Western subspecies has a whitish belly; eastern subspecies has a bright yellow belly. They eat insects in trees and on ground. Nest a grassy cup placed on ground to 10 ft. high in conifers. **Voice:** Song a buzzy trill changing in pitch; call a sharp *tsup*.

Winter (Western Palm)

Breeding Info — E: 4–5 **I:** 12 **N:** 12 **B:** 1–2

Male

American Redstart

Setophaga ruticilla Length: 5" Wingspan: 7¾"

The male American Redstart is boldly colored, with black upperparts and orange patches on the sides of his breast and tail and at the base of his wings. The female, equally beautiful but subtler, has olive upperparts, a gray head, and yellow patches on the sides of her breast and tail and at the base of her wings. Males in their 1st year look like adult females but with some blotches of black on the face or breast. Redstarts often feed by sallying out to catch aerial insects. Nest a cup of grasses bound with spider silk and placed in a small tree or shrub. **Voice:** Song a series of whistles with an accented downslurred ending; call a sweet *chip*.

Female

Breeding Info — **E**: 2–5 **I**: 12 **N**: 8–9 **B**: 1

Yellow-throated Warbler
Setophaga dominica Length: 5¼" Wingspan: 8"

The Yellow-throated Warbler makes its identification easy, for the adults have the same
plumage all year and the sexes look alike — there
is only one plumage to identify! It is recognized
by its bright yellow throat, black mask, and black
streaking on the flanks. The Yellow-throated
breeds primarily in the Southeast and overwinters in Florida and along the Gulf Coast. It eats
insects, spiders, and even small lizards; may
come to bird feeders for seeds and suet. Nest of
grasses and plant down bound with spider silk is
placed on a pine limb or in hanging Spanish moss.
Voice: Song a descending set of whistles ending
on a higher note; call a musical *tsip*.

Breeding Info — E: 4–5 **I**: 12–13 **N**: 9–11 **B**: 1–2

Summer male

Chestnut-sided Warbler

Setophaga pensylvanica Length: 5" Wingspan: 7¾"

Moist sunny shrublands are the habitat of the Chestnut-sided Warbler. Its explosive little song sounds like "Pleased pleased pleased to meet-cha." It is a colorful warbler with distinctive swaths of chestnut on its flanks, more so on the male than the female. In winter it loses its black facial marks and becomes a simplified olive-green above with less chestnut on the flanks. Chestnut-sideds eat insects, spiders, seeds, and berries. Nest of weeds, grape bark, and plant down placed in shrubs or brambles. **Voice:** Call is a sweet *chip*.

Breeding Info – **E**: 3–5 **I**: 12–14 **N**: 10–12 **B**: 1

Winter

Male

Summary Tanager

Piranga rubra Length: 7¾" Wingspan: 12"

This is the most southern of the tanagers and
easy to recognize — the male is all red and the
female olive-yellow (no black wings and tail as
on the Scarlet Tanager, no wingbars as on the
Western Tanager). The only variation is on the
1st-year male, which is a mixture of red and olive.
It lives in deciduous forests and pine-oak woodlands, where it feeds on insects and fruit. It also
may eat bees and even raid their nests. Sometimes comes to feeders for fruit. Nest of weed
stems, bark strips, and grass placed on a horizontal limb 10–35 ft. high. **Voice:** Song a series of
2-part burry whistled phrases like *jurit jeroo jaree
jeray*; call a distinctive *kituk* or *kitituk*.

Breeding Info—E: 3–5 **I:** 12 **N:** 10–11 **B:** 1

Female

Male

Scarlet Tanager
Piranga olivacea Length: 7" Wingspan: 11½"

Despite the male's beautiful red color, he may be hard to spot when hidden by treetop foliage. This is a summer bird mainly of northeastern and midwestern deciduous forests. The summer male is our only all-red bird with black wings; in winter it is mostly yellow with black wings. The female, all year, is greenish-yellow overall with unmarked dark brown wings. Seen in Gulf states in spring and fall as it migrates to and from Central America. Eats insects and berries. Cup-like nest of twigs, rootlets, and grasses placed in trees 5–75 ft. high. **Voice:** Song is a series of hoarse whistles, call a distinctive *chip-burr*.

Breeding Info — **E**: 2–5 **I**: 12–14 **N**: 9–10 **B**: 1

Winter male

Male

Western Tanager
Piranga ludoviciana Length: 7¼" Wingspan: 11½"

The Western is our only common tanager with wingbars. The summer male has a yellow body, black back and wings, and a reddish face; the winter male has an all-yellow face and a greenish back. The female, year-round, is mostly grayish-yellow overall with white wingbars. She is best recognized by her "tanager" bill, short, blunt-tipped, deep-based, and with strongly curved top to upper bill. Forages for insects in the tree canopy; may catch wasps in midair; comes to feeders for orange halves. Nest of twigs and rootlets placed near tips of branches.
Voice: A series of burry, whistled phrases like *jurit jeroo jaree jerilay*; calls include a *chideet* or *chidereet* and a rising *weeyeet*.

Breeding Info—**E**: 3–5 **I**: 13 **N**: 13–15 **B**: 1

Female

Male

Northern Cardinal
Cardinalis cardinalis Length: 8¾" Wingspan: 12"

The Northern Cardinal is undoubtedly one of America's most popular birds. The bright red of the male and the close relationship between the pair endear them to everyone's heart. In courtship the male picks up a seed, flies to the female, and places the seed in her bill. Cardinals are easily attracted to feeders, especially for sunflower seeds, and they will nest in backyards that have some dense shrubbery. Nest of twigs, bark strips, and leaves lined with fine grass and hair is placed in a tree or shrub 1–15 ft. high. **Voice:** Song is a series of slurred whistles like *woit woit woit cheer cheer cheer*; calls include a metallic *chip* and a harsh *kwut*.

Female

Breeding Info — E: 2–5 **I**: 12–13 **N**: 9–11 **B**: 1–4

Male

Rose-breasted Grosbeak

Pheucticus ludovicianus   Length: 8¼"
Wingspan: 12½"

The male Rose-breasted Grosbeak certainly grabs
the attention when he shows up at bird feeders,
with his jet-black head, pure white belly, and
shocking splash of red across his breast. If this were
not enough, in flight he reveals white patches on
his upperwings and pink underwings. The female
is mostly streaked brown and white. Grosbeaks are
medium-sized birds with large heads and distinctive short deep-based bills. They eat insects, tree
seeds and buds, fruit, and sunflower seeds at
feeders. Nest of twigs lined with rootlets placed
in a tree 5–25 ft. high. **Voice:** An extended rapid
series of whistles, "like a robin in a hurry"; calls
include a squeak like a sneaker on a gym floor.

Female

Breeding Info — **E**: 3–6 **I**: 12–14 **N**: 9–12 **B**: 1–2

Male

Black-headed Grosbeak
Pheucticus melanocephalus Length: 8¼"
Wingspan: 12½"

This is the popular grosbeak of the West, being found in a variety of habitats from open woods to river edges and from mountain canyons to parks and gardens. The male has a black head and back, and an orangish body; the female is similar but with a brownish head with a white eyebrow. Feeds mostly in tree canopies, eating insects, tree seeds and buds, and fruits; will come to feeders for mixed seeds and fruit. Nest of twigs and weed stems, lined with rootlets placed in a tree or shrub 4–25 ft. high. **Voice:** A series of whistled notes; calls include a sharp *chik*.

Breeding Info—**E**: 2–5 **I**: 12–13 **N**: 11–12 **B**: 1

Female

Male

Blue Grosbeak
Passerina caerulea Length: 6¾" Wingspan: 11"

The male Blue Grosbeak is an incredible deep
blue and the female a lovely warm brown. The
key to distinguishing them from their look-alike
cousin, the Indigo Bunting, is to look for the
grosbeak's two prominent wide rusty wingbars.
These birds breed mainly across the southern
half of the United States and then leave for the
winter. During migration they may be found in
substantial flocks, sometimes along with Indigo
Buntings, their smaller relative. They eat seeds,
insects, and spiders on or near the ground. Nest
of rootlets, grasses, cotton, and sometimes
snakeskins is placed in shrubs or trees 3–12 ft.
high. **Voice:** Song a long series of burry whistles;
call a sharp *chink*.

Female

Breeding Info — **E**: 2–5 **I**: 11–12 **N**: 9–13 **B**: 1–2

Male

Indigo Bunting

Passerina cyanea Length: 5½" Wingspan: 8"

In an overgrown farm field, a meadow's tangled edge, or among roadside wildflowers, this is where you are likely to hear the sweet, paired, whistled notes of the Indigo Bunting. These early succession habitats are its favorite for breeding; it places its cuplike nest of grasses in the fork of a young sapling. The male is an unmistakable all blue with a short conical bill. The female, however, is conspicuous for her lack of identification clues; she is all pale brown with a hint of streaking below and sometimes a hint of blue on wings, tail, and rump. Indigos eat insects, seeds, and berries; will come to feeders for finch seeds like white millet. **Voice:** Calls include a sharp *tsik* and a buzzy *dzeet*.

Female

Breeding Info—**E**: 2–6 **I**: 12 **N**: 10–12 **B**: 1–2

Male

Lazuli Bunting

Passerina amoena Length: 5½" Wingspan: 8¾"

The Lazuli is the main bunting of the West, choosing brushy areas and streamsides as its preferred habitat. The male has an electric-blue head and back contrasting with an orangish breast and flanks; its broad white wingbars are an additional helpful clue. The female is a plain pale brown with two thin buffy wingbars and a tint of blue on her rump and wings. Lazuli Buntings feed on or near the ground, eating insects and seeds; will come to feeders. Nest of grasses lined with finer grasses and horsehair placed in shrub 1–10 ft. above ground. **Voice:** Song a long series of high buzzy whistles; calls a sharp *pik* and buzzy *dzeeet*.

Female

Breeding Info—E: 3–5 **I:** 12 **N:** 10–15 **B:** 1–3

Male

Painted Bunting

Passerina ciris Length: 5½" Wingspan: 8¾"

Every birdwatcher traveling in the South wants
to see the male Painted Bunting – undoubtedly
the most richly colored bird in North America.
The female has lovely bright green tones on her
body. Painted Buntings summer in Texas and
adjoining states and along the southeastern and
Florida coasts. They winter in southern Florida,
where they often come to feeders. Brushy edges,
roadsides, and gardens are favorite haunts for
foraging near or on the ground for seeds and
insects. Nest of grasses and leaves lined with hair
placed in shrub or tree 3–25 ft. high. **Voice:** Song
a rapid series of varied whistles; calls a sharp *chip*
and thin *tseeet*.

Breeding Info – **E:** 3–5 **I:** 11–12 **N:** 12–14 **B:** 1–4

Female

Winter

Snow Bunting

Plectrophenax nivalis Length: 6¾" Wingspan: 14"

The Snow Bunting is primarily a winter visitor
to the northern states and southern Canada,
coming to us after breeding in the northernmost
tundra. It can be found as flocks in barren areas,
eating seeds from the sparse vegetation. The
birds typically feed for a bit and then fly up in a
swirl revealing their boldly patterned black-and-
white wings before settling down again. Snow
Buntings in winter are mostly white with a buffy
wash on the face, nape, and sides of breast. Nest
of moss, grasses, and earth placed on ground in
rocky area. **Voice:** Calls include a rapid *chidit*, a
harsh *djeeet*, and a whistled *tew*.

Breeding Info — E: 3–9 **I:** 10–16 **N:** 10–17 **B:** 1–2

Winter

Male

Eastern Towhee

Pipilo erythrophthalmus Length: 8½"
Wingspan: 10½"

Towhees live in the open shrubby understory
of deciduous forests or pines. Their calls and
songs are loud and memorable. While feeding,
towhees jump forward, then back, raking leaves
and pine needles to expose insects, spiders,
seeds, and fallen berries. This noisy shuffling
may help you locate the bird. Comes to feeders
for seed on the ground. The male has a black
hood and upperparts, and rufous flanks beside a
white belly. The female is similar but replaces the
black with rich brown. Nest of leaves and grasses
placed on the ground in a scratched depression
under shrubbery. **Voice:** Song much like the
phrase "Drink your teeee"; call a loud *chewink*.

Female

Breeding Info — **E**: 2–6 **I**: 12–13 **N**: 10–12 **B**: 1–3

Male

Spotted Towhee

Pipilo maculatus Length: 8½" Wingspan: 10½"

The Spotted Towhee is the western counterpart of the Eastern Towhee, having all the same feeding and nesting habits. In fact, the two were previously considered one species. The difference lies in the plumage and distinction between the sexes. The male Spotted has variable white spotting on his black back and wings. In the Spotted, the female and male are very similar, the female being dark grayish-brown to blackish where the male is jet black. **Voice:** Song is one or more introductory notes followed by a harsh trill. Calls include a harsh upslurred *jaweee*.

Breeding Info — **E:** 2–6 **I:** 12–13 **N:** 10–12 **B:** 1–3

Female

California Towhee

Melozone crissalis Length: 9" Wingspan: 11½"

This rather large towhee is found in dense shrubs, streamsides, and gardens up and down the western half of California and a little bit into Oregon. It readily comes to feeders for mixed seed scattered on the ground. A rather plain grayish-brown overall except for its orangish undertail and suggestion of orange on its face. It is similar to the Canyon Towhee of the Southwest but separated by range. Nest of small twigs, grasses, plant stems placed on the ground or up to 12 ft. high in shrubs. **Voice:** Song a halting series of metallic chips; calls include a *tssp*, metallic *chink*, and buzzy *dzeee*.

Breeding Info — **E**: 2–6 **I**: 11 **N**: 8 **B**: 2–3

Adult summer

Chipping Sparrow

Spizella passerina Length: 5½" Wingspan: 8½"

This is a wonderful dooryard sparrow, happy to live near suburban and country houses, taking advantage of feeders and boldly nesting in dense evergreens of foundation plantings. The Chipping Sparrow is clear-breasted and has a distinctive rusty cap in summer; in winter this is replaced by a brown cap finely streaked with black. When not at bird feeders, it feeds on the ground, eating seeds from weeds and grasses. The firm cuplike nest is made of grasses and lined with horsehair or similar material; usually placed 3–10 ft. high. **Voice:** Song is a long chipping trill, fairly dry and harsh in tone.

Breeding Info — **E**: 3–4 **I**: 11–12 **N**: 7–10 **B**: 1–2

Adult winter

American Tree Sparrow

Spizella arborea Length: 6¼" Wingspan: 9½"

Sighting an American Tree Sparrow is usually an indication that winter is here, for this species breeds farther north than most other sparrows and comes down into the United States in winter. This is a relatively large colorful sparrow with a bright rusty crown, eyeline, wings, and back, and two white wingbars. The breast is clear and unmarked gray except for a central dark spot. Tree Sparrows love to feed on the seeds of shrubs or in weedy fields; they also come to feeders for mixed seeds. Nest is cuplike, made of grasses and bark strips placed on the ground or in a shrub. **Voice:** Song is a short phrase of sweet whistles and warbles; call is a tinkling *teeahleet*.

Breeding Info – **E**: 3–5 **I**: 12–13 **N**: 9–10 **B**: 1

Savannah Sparrow

Passerculus sandwichensis Length: 5½"
Wingspan: 6¾"

The Savannah is a heavily streaked sparrow, generally with some variable yellow in the eyebrow.
Streaking on the breast is fine and dark, and only
occasionally coalesces into a central dot. The
preferred nesting habitat for the Savannah is a
large expanse of grassy meadows or agricultural
fields. Here the male perches atop a weed stem
or tall grass blade and gives his soft insectlike
song of a few harsh introductory *chip*s followed
by 2 buzzy high-pitched trills like *ts ts ts tseeee
chaaay*. Savannahs feed on the ground, often
running about as they look for seeds and insects.
Cuplike nest of grasses and moss placed on the
ground. **Voice:** Calls include a high *tsip*.

Breeding Info – **E:** 4–5 **I:** 12 **N:** 14 **B:** 1–2

Eastern

Fox Sparrow

Passerella iliaca Length: 7" Wingspan: 10½"

The Fox Sparrow is large and dramatically colored, with bright reddish-brown upperparts and heavily streaked underparts. It likes shrubby areas and gardens, where it feeds on the ground, jumping forward and raking back to expose insects and seeds. It winters mainly in the Southeast and along the West Coast, and thus often shows up at feeders elsewhere while on migration in fall and spring. Fox Sparrows can look different: eastern birds are more reddish or "foxy"; western birds are duller brown or with gray tones. Nest of grasses and leaves lined with hair is placed on the ground under a shrub. **Voice:** Song a series of rich whistles given at a leisurely pace.

Eastern

Breeding Info – E: 4–6 **I:** 11–14 **N:** 7–12 **B:** 1–2

Swamp Sparrow
Melospiza georgiana Length: 5¾" Wingspan: 7¼"

The Swamp Sparrow is common but easily over-looked because it lives in marshes, bogs, and the shrubby edges of lakes and rivers. The best way to locate the bird is to hear its song, which is a musical trill on one note coming from the male as he perches on cattails or other marsh vegetation. Swamp Sparrows winter mostly in the Southeast and are generally quieter than in summer, but can be recognized by their chestnut cap, reddish-brown wings, and dingy underparts. They feed on the ground or in very shallow water, eating insects and seeds. Bulky cuplike nest of grasses is placed in cattails or shrubs to 2 ft. high. **Voice:** Calls include a short *tsit* and a soft *chip*.

Breeding Info — **E**: 4–5 **I**: 12–15 **N**: 10–13 **B**: 1–2

Song Sparrow

Melospiza melodia Length: 4¾"–6¾"
Wingspan: 8¼"

The Song Sparrow is the quintessential sparrow, heavily streaked brown above and below and with a central breast dot. The male perches atop a shrub to deliver a long melodic song that sounds a lot like "Maids, maids, maids, put on your teakettle kettle kettle." What a welcome song to hear in the early spring garden! At home in the wilds or backyards, it eats seeds, insects, some fruit, and may forage for seeds under feeders. It is a widespread breeder in North America and overwinters throughout most of the lower 48 states. Nest of grasses placed on the ground or in a shrub 1–4 ft. high. **Voice:** Calls include a soft *tsip*, a high *tseeet*, and a louder deeper *tchup*.

Breeding Info — **E:** 3–5 **I:** 12–13 **N:** 10 **B:** 1–3

White morph

White-throated Sparrow

Zonotrichia albicollis Length: 6¾" Wingspan: 9"

The White-throated Sparrow has one of the more beautiful songs of North American birds. It is a lovely series of 3 to 7 long whistles, the first usually lower pitched, the later ones often quavering. It has been likened to the words "Ole Sam Peabody Peabody Peabody." It is a defining sound of the northern coniferous woods. You can recognize White-throats by the white throat and the yellow dot in front of the eye. They come in two morphs but are the same species: the white morph with white and black crown stripes, and the tan morph with brown and tan crown stripes. White-throats eat seeds and insects, and may forage under seed feeders. Nest of grasses lined with hair is placed on the ground. **Voice:** Calls include a drawn-out *tseeet* and a high *pink*.

Tan morph

Breeding Info—**E**: 4–6 **I**: 11–14 **N**: 7–12 **B**: 1–2

Adult

White-crowned Sparrow

Zonotrichia leucophrys Length: 7" Wingspan: 9½"

The bold black and white head stripes and clear gray face and breast are the best way to identify the handsome White-crowned Sparrow. In the East, it is seen mostly on migration and in winter; on the West Coast, it can be seen year-round. 1st-winter birds have brown crown stripes. Feeds on ground by raking back leaf litter to find seeds and insects; also comes to seed under feeders. Bulky nest of grasses lined with finer materials placed on ground or in small shrub or tree. **Voice:** Song of long whistles and short warbles; calls include *pink* and rising *seeyeet*.

Breeding Info—E: 3–5 **I:** 11–15 **N:** 10 **B:** 1–4

1st winter

Adult

Golden-crowned Sparrow

Zonotrichia atricapilla Length: 7¼"
Wingspan: 9½"

This is a large sparrow of the West Coast, breeding from British Columbia through Alaska and
wintering from Washington to southern California. It is similar to the White-crowned Sparrow
but has a broad black band surrounding a yellow
crown. In winter the crown is brown with a duller
yellow center. Feeds on ground, eating seeds,
insects, fruit, also buds, leaves, and flowers, and
eats seeds under feeders. Large nest of grasses,
ferns, and sticks placed in a depression on the
ground. **Voice:** A series of 3 clear whistled notes
descending in pitch, like "Oh dear me"; calls
include a high *tseeep*.

1st winter

Breeding Info — E: 3–5 I: 11–14 N: 9–11 B: 1–2

Slate-colored

Pink-sided

Gray-headed

Oregon

Dark-eyed Junco

Junco hyemalis Length: 6¼" Wingspan: 9¼"

The Dark-eyed Junco is one of the most common winter feeder birds in North America. It is attracted to mixed seed placed in a tray or sprinkled on the ground. It is the "snowbird" breeding in the North and at higher elevations, then flying south and to lower elevations to spend the winter. Some liken its coloration to a winter day—gray skies above, white snow below. Juncos in the West may look different and be dark-hooded (Oregon Junco), Pink-sided, or Gray-headed. Juncos eat seeds on the ground. Nest of grasses, moss, and pine needles is in a depression on the ground. **Voice:** Song a musical trill on one note; calls include a buzzy *zeeet*, a sharp *tik*, and a *tewtewtew*.

Slate-colored

Breeding Info—E: 3–6 **I**: 12–13 **N**: 9–13 **B**: 1–2

Male

Red-winged Blackbird

Agelaius phoeniceus Length: 8¾" Wingspan: 13"

Red-winged Blackbirds are the signature birds of swamplands across North America. The male's strident song of *okalee* is given as he sways atop a cattail stalk. His red shoulder boldly bordered by yellow can be hidden or fully exposed as he patrols his territory. The female is camouflaged with strong brown streaking but makes her presence known through her own unique call of *ch'ch'ch'chee chee*. Redwings feed on seeds and insects, and come to feeders for mixed seed and cracked corn. Swamps are their preferred habitat, but they also nest in lush fields. Nest of reeds and grasses in cattails and shrubs 3–8 ft. high. **Voice:** Calls include a downslurred *tseeyeer* and a harsh *chek*.

Female

Breeding Info – E: 3–5 **I**: 11 **N**: 11 **B**: 1–3

Male

Yellow-headed Blackbird
Xanthocephalus xanthocephalus Length: 9½"
Wingspan: 15"

This is a large and dramatic blackbird of cattail
marshes and prairie wetlands. The ratchety song
of the male sounds a little like a Red-winged
Blackbird with a sore throat. In addition to the
obvious yellow head and black body, the male
can reveal bold white "shoulder" patches in
flight and during territorial displays. The female
is smaller, and dark brown with a variably yellow
breast bordered by some white streaking. They
feed on insects and seeds in marshes and agricultural fields. Nest of reeds and grasses placed in
vegetation up to 7 ft. high. **Voice:** Song a gargled
low-pitched *gugleko graaah*; calls include a 2-part
chukuk and a *chek*.

Female

Breeding Info — **E**: 3–5 **I**: 11–13 **N**: 9–12 **B**: 1–2

Eastern

Western

Eastern/Western Meadowlark

Sturnella magna (Eastern)
Sturnella neglecta (Western)
Length: 9½"   Wingspan: 14"

Typically, a Meadowlark is found in open farmland, on a fence post, singing. Although it has
bright yellow underparts and a black V on its
breast, these can be hidden when the bird is
facing away. Short flights reveal shallow wingbeats and white outer tail feathers. These two
species are almost identical, except the Western
has more yellow on its cheek and less white on its
tail. Feeds on ground on insects and seeds. Nest
is domed, made of grasses, placed on the ground
in a slight depression. **Voice:** Song of both species
starts with a few clear whistles; that of Western
ends in a warble. Calls include a *dzeeert* for Eastern and high-pitched *weeet* for Western.

Eastern

Western

Breeding Info — E: 3–7 **I:** 13–15 **N:** 11–12 **B:** 1–2

Male summer

Bobolink

Dolichonyx oryzivorus Length: 7" Wingspan: 11½"

Expansive hayfields are the Bobolink's favorite habitat. Here the striking black-and-white male flies up from the grasses and during a slow and stalling descent gives his marvelous jumbled song of warbles and harsh notes. In summer the male is black with a white back and yellowish nape; the female has a warm buffy look, a brown crown with whitish central stripe, and a streaked brown back. In fall the male molts into plumage like the female's. Bobolinks eat insects and seeds on the ground and cultivated rice crops in the South. Nest of grasses placed on ground. **Voice:** Calls include a descending *seeyew*, a harsh *chek*, and a *pink* given in flight after breeding.

Breeding Info—**E**: 4–6 **I**: 12 **N**: 10–14 **B**: 1

Female

Male

Common Grackle

Quiscalus quiscula Length: 12½" Wingspan: 17"

The Common Grackle is our most widespread grackle in the East and Midwest. It is medium-sized, with a long tail and a hefty, deep-based, slightly downcurved bill. Both sexes of Common Grackles have bluish iridescent hoods and yellow eyes. The male's body and wings are a contrasting bronze or matching iridescent blue; the female's body and wings are blackish brown with little iridescence. Grackles eat seeds, insects, and small animals on the ground or in shallow water; come to feeders for mixed seed. Bulky nest of grass, twigs, and mud in a tree or shrub 3–30 ft. above ground or over water. **Voice:** Song a screechy *screedleeek*; calls include *chak* and a drawn-out *chaaah*.

Female

Breeding Info—**E**: 4–7 **I**: 13–14 **N**: 12–16 **B**: 1

Male

Boat-tailed Grackle

Quiscalus major Length: 14½" –16½"
Wingspan: 17½"–23"

This is the coastal grackle, living along the
Atlantic and Gulf coasts from New England to
Texas. It is larger than the Common Grackle, has
a substantially longer tail, and longer legs. The
female looks different from the male, having
a pale brown head and dark brown body; the
male is glossy black overall. Either sex can have
a yellow or brown eye (depending on the subspecies). During display flights, all male grackles
hold their long tails in a distinctive V; females do
not. Grackles are often seen in groups, feeding
on scraps in parks or parking lots. Bulky nest of
grasses and mud or cow dung in shrub or tree
3–12 ft. high. **Voice:** Song a scratchy *b'jee b'jee
b'jee*; call a *tseet* or *jeer jeer*.

Female

Breeding Info—**E**: 3–5 **I**: 13–15 **N**: 20–23 **B**: 1–3

Male

Great-tailed Grackle

Quiscalus mexicanus Length: 15"–18"
Wingspan: 19"–23"

The Great-tailed is the main grackle of the
Southwest. It is our largest grackle, with a very
long tail and proportionately rather small head.
The male is blackish with violet-blue iridescence
overall; the female is pale to warm brown on
head and body with darker brown wings and
tail. Great-taileds feed on insects, grain, and
small fish from parks to farmlands. Almost
always in flocks—nesting in colonies and forming
huge winter roosts. Nest of grasses and mud
or cow dung placed in shrub or tree. **Voice:**
Song a strange mixture of mechanical grating,
harsh sounds, and squeals; calls include a rising
tooweeet, a chatter, and a low *chuk*.

Female

Breeding Info—E: 3–4 I: 13–14 N: 20–23 B: 1–2

Male

Brewer's Blackbird

Euphagus cyanocephalus Length: 9"
Wingspan: 15½"

The Brewer's is a wide-ranging blackbird of the
West, usually encountered as flocks in open
country, from urban areas to farmlands. It feeds
on the ground, eating insects, seeds, and berries.
In winter it may move about in mixed flocks with
other blackbirds and grackles; on winter nights
in communal roosts in marshes it may number in
the millions. The male is black with glossy iridescence and a yellow eye; the female is pale to dark
gray overall and with usually a dark eye. Nest
of twigs, grass, and mud or cow dung placed in
vegetation on or just above the ground. **Voice:**
Song a short scratchy *krt jee*; calls include a *chek*,
a *tseeyur*, and a chattering *ch'ch'ch'ch*.

Female

Breeding Info — **E:** 3–7 **I:** 12–14 **N:** 13–14 **B:** 1–2

Male

Brown-headed Cowbird
Molothrus ater Length: 7½" Wingspan: 12"

The cowbird lays its eggs exclusively in other birds' nests and lets them raise its young. This can reduce the breeding success of the pairs they parasitize. That said, 97 percent of cowbird eggs and nestlings fail to reach adulthood, often because the host pair tosses out the cowbird egg. The male cowbird is glossy black with a brown head; the female grayish-brown with a whitish chin and black bill. Feeds on the ground, eating weed seeds, grain, insects; may follow farm animals, eating insects they stir up. **Voice:** Song a gurgling sound followed by a squeaky whistle, like *bublocomtseee*; calls include a double whistle like *pseeeseee*, a *chuk*, and a chattering *ch'ch'ch*.

Female

Breeding Info—E: ? I: 10–13 N: 9–11 B: ?

Male

Orchard Oriole

Icterus spurius Length: 7¼" Wingspan: 9½"

Compared to other orioles, the Orchard is smaller, with a shorter squared-off tail and short slightly downcurved bill. The male is very dark with a black hood, back, wings, and tail; his belly and rump are chestnut. The female is greenish-yellow with slightly grayer back. 1st-year males look like adult females but have a black throat. Orchard Orioles live in orchards, open woods, and parks. They eat insects, fruit, and tree blossoms in the canopy; may come to hummingbird feeders. Pouchlike nest of woven grass blades suspended from horizontal branch 12–20 ft. high. **Voice:** Song of varied whistles and buzzes; calls include chattering, a downslurred *teew*, and a *ch'peet*.

Female

Breeding Info—**E**: 3–7 **I**: 12–14 **N**: 11–14 **B**: 1

Male

Baltimore Oriole

Icterus galbula Length: 8¾" Wingspan: 11½"

Besides being a baseball team symbol, the Baltimore Oriole is exciting in its own right. The male's striking plumage, his loud whistled song, the female's pendant nest, and their feeding on oranges make them welcome in any backyard. The male's black hood and bright orange body are unmistakable; the female is similar, but her head and back are mottled and her central tail feathers dusky orange (those of the male are black). Live in open deciduous woods; eat insects, fruit, and flower nectar, and come to feeders for oranges and sugar water. Pendent woven nest of long plant fibers suspended from branch 6–60 ft. high. **Voice:** Song is several varied clear whistles; calls a chattering and an upslurred *weeet*.

Breeding Info – **E:** 4–6 **I:** 12–14 **N:** 12–14 **B:** 1

Female

Male

Bullock's Oriole

Icterus bullockii Length: 9" Wingspan: 12"

The Bullock's is the most widespread oriole of the West, often suspending its long woven nest from the tips of streamside cottonwoods. In this environment of dappled light, you are likely to hear its short fairly chattering song. The male is black above and orange below with a black eyeline and broad white wingbar; the female has an orangish head and breast, pale gray belly, and two white wingbars. The 1st-year male is like the female but with a centrally black chin and throat. Bullock's eat insects, fruit, and flower nectar. Woven nest of long plant fibers suspended from branch 6–60 ft. high. **Voice:** Calls include a *chet*, a *weet*, and a rough chattering.

Female

Breeding Info – **E**: 4–6 **I**: 12–14 **N**: 12–14 **B**: 1

Male

Purple Finch

Haemorhous purpureus Length: 6" Wingspan: 10"

This is a colorful finch mostly of northern coniferous forests, but it also visits edge habitats, gardens, orchards, and backyard feeders. The male looks like he was dipped in raspberry juice and has little to no brown streaking on the flanks; the female is coarsely streaked brown all over and has a broad white eyebrow. They forage from on the ground to high in trees eating insects, seeds, and fruits, and coming to feeders for sunflower and millet. Nest of twigs, grasses, and rootlets placed in a tree 5–60 ft. high. **Voice:** Song a rapid, short, musical warbling; calls include a sharp flat *pik* and a whistled *tweeoo*.

Breeding Info – **E**: 36 **I**: 13 **N**: 14 **B**: 1–2

Female

Male

House Finch

Haemorhous mexicanus Length: 6" Wingspan: 9½"

This is a widespread little finch of backyards, urban parks, farmlands, and canyons. The male, with his rosy head and breast, adds a touch of color at bird feeders where this species comes to eat mixed seeds. House Finches nest just about anywhere and often near homes, such as in foundation plantings, porch vines, hanging planters, and birdhouses. The male has variable orangish red on head, breast, and rump; his flanks are strongly streaked with brown. The female is pale brown, strongly streaked beneath, and finely streaked on face. Nest of twigs, grasses, leaves. **Voice:** Song a long rapid warble; calls include a *ch'weet.*

Female

Breeding Info— E: 2–6 **I:** 12–16 **N:** 11–19 **B:** 1–3

Male

Common Redpoll

Acanthis flammea Length: 5¼" Wingspan: 9"

Common Redpolls are usually just a winter visitor, forsaking their Arctic breeding grounds to enjoy bird feeders in southern Canada. In some years, large numbers may visit the lower 48 states when their predominant food of tree seeds is scarce. Look for a deep-chested little bird with a short stubby bill; they are streaked brown with a red forehead and black chin. The male has pink on his chest and flanks; the female has little or none. When not at your feeders, they are perched atop birches and alders eating their tiny seeds. Nest of grasses and twigs placed in bush or on ground. **Voice:** Song a long series of short calls and trills; calls include a *jeeeyeeet* and *chew chew chew*.

Female

Breeding Info — **E:** 4–7 **I:** 10–11 **N:** 12 **B:** 1

Male

Evening Grosbeak

Coccothraustes vespertinus Length: 8"
Wingspan: 14"

The sleighbell-like calls of a flock of Evening
Grosbeaks is the perfect accompaniment to
a cold winter day in the holiday season. They
breed in northern climes and higher elevations
but can disperse far to the south in winter. The
male looks like a space cadet with his darkish
hood and bright yellow eyebrow; the female is a
more subtle combination of grays and yellows,
with black-and-white wings and tail. Their huge
bill is pale green in summer, horn-colored in winter. Evening Grosbeaks feed mainly on tree seeds
such as maple and box elder; come to feeders for
sunflower seeds. Nest of lichens, twigs, and rootlets placed in tree 20–100 ft. high. **Voice:** Song a
halting warble; calls include a repeated *cleer cleer.*

Female

Breeding Info — **E:** 2–5 **I:** 11–14 **N:** 13–14 **B:** 1–2

Winter

Male summer

American Goldfinch

Spinus tristis Length: 5" Wingspan: 9"

The American Goldfinch is a highly desired species at the bird feeder, adding a touch of color at all times of year. The summer male is bright yellow with a black cap; the female is all yellow with brownish wings. In winter both sexes look like the female, but with muted colors. In August look for goldfinches with buffy wingbars; these are fledglings. Goldfinches eat seeds from trees and wildflowers, and take sunflower and thistle at feeders. Nest of downy plant materials bound with caterpillar silk placed in shrub or small tree 4–20 ft. high. **Voice:** Song a continuous warbling for 30 seconds or more; calls include a rising *seeyeet*, a *titewtewtew*, and a *bearbee bearbeebee*.

Female summer

Breeding Info — **E**: 3–7 **I**: 12–14 **N**: 11–15 **B**: 1–2

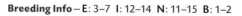

Male

Male

Lesser Goldfinch

Spinus psaltria Length: 4½" Wingspan: 8"

This goldfinch of the West comes in two subspecies: in one, the male is all black above
and all yellow below; in the other, the male is
greenish-yellow overall with a black cap, wings,
and tail. The former is in the Southwest, the
latter in the West. In both, the female looks the
same, with bright to dull yellow underparts,
olive upperparts, and a grayish bill. Can live in
woods, edges, and gardens, where it eats seeds,
flower buds, and berries. Comes to feeders for
sunflower and thistle. Nest of bark, moss, and
plant stems placed in shrub or tree 2–30 ft. high.
Voice: Song a series of chips and whistles; calls
include *bearbee* and *chit chit chit*.

Female

Breeding Info — E: 3–6 **I:** 12 **N:** 10–12 **B:** 1–2

Pine Siskin

Spinus pinus Length: 5" Wingspan: 9"

The Pine Siskin is most often seen in winter at feeders, where it prefers to eat thistle and sunflower seeds. It breeds in the north and at higher elevations and winters south throughout the States. It is a small slim bird with a short well-notched tail and slender fine-pointed bill. It is heavily streaked brown with variable touches of yellow on the wings and tail. It feeds on the ground and forages in foliage for conifer seeds, weed seeds, flower buds, and insects; in some years large numbers may come to feeders. Nest of grasses, twigs, and rootlets placed in tree 3–50 ft. high. **Voice:** Song a series of harsh chirps; calls include a buzzy rising *dzzzzeeee* and a flat *tseeyeet*.

Breeding Info — **E**: 1–5 **I**: 13 **N**: 14–15 **B**: 1–2

Male

House Sparrow

Passer domesticus Length: 6¼" Wingspan: 9½"

This is the ultimate city bird, introduced into North America in the mid-1800s and now widespread. You cannot go far on a city sidewalk without encountering a House Sparrow. They have no trouble making a living on scraps from humans and building their nests in the many crannies city structures offer—bridges, signs, store awnings, and traffic lights. The male has rusty brown upperparts, a gray crown and rump, and a black chin and breast; the female is dull brown with a streaked back and plain face. Nest of grasses, feathers, cloth, and trash placed in nook. It can take over birdhouses from native species. **Voice:** Song a series of 2-part chirps like *chirup chireep chirup*; calls include a churring *ch'ch'ch* and a 2-part *kwer kwer*.

Female

Breeding Info—**E**: 3–7 **I**: 10–14 **N**: 14–17 **B**: 1–3

Photo Credits

All photos by Lillian Stokes except those listed below.

Numbers refer to pages; letters refer to position on the page. Abbreviations are as follows:

M = main large photo
ML = main left large photo
MR = main right large photo
MT = main top large photo
MB = main bottom large photo
MTR = main top right photo
MBL = main bottom left photo
MBR = main bottom right photo
I = small photo insert within large photo
C = small photo in lower right corner

Clair Postmus
4M, 4I, 8MT, 8MB, 15MT, 15MB, 19MT, 19C, 20MT, 20MB, 20MC, 50C, 58C, 77M, 77I, 78M, 81M, 87C, 96C, 107M, 107C, 112M, 112C, 117M, 120M, 120C, 121M, 122C, 123C, 130M, 130C, 131M, 131C, 132M, 132C, 133M, 133C, 134M, 134C, 140M, 145C, 146M, 146C, 148M, 148C, 161M, 161C, 166M, 166C, 180C, 187M, 187C, 188M, 188C, 195M, 195C, 208C, 216C, 218M, 221M, 221C, 222M, 224M, 224C, 228M, 228C, 239MTR, 239MBL, 239MBR, 241M, 241C, 242MR, 246M, 247M, 247C, 251M, 251C, 257M, 257I, 257C

Bob Steele
18MT, 18MB, 18C, 19MB, 66M, 66C, 78I, 78C, 81I, 81C, 96M, 104M, 104C, 111M, 111C, 114M, 114C, 125M, 125C, 136M, 136C, 140C, 158M, 158C, 160M, 160C, 170M, 170C, 173M, 173C, 174M, 174C, 176M, 176C, 180M, 184M, 206C, 208M, 218C, 229M, 229C, 238M, 238C, 242ML, 246C

Alan Murphy
108C, 138M, 205C, 228C, 249M, 249C

Jim Zipp
99C, 123M, 213C

Index

A

Acanthis flammea, 254
Accipiter cooperii, 98
— *striatus*, 99
Actitis macularia, 58
Aeronautes saxatalis, 136
Agelaius phoeniceus, 240
Aix sponsa, 8
Anas acuta, 15
— *americana*, 9
— *clypeata*, 14
— *crecca*, 16
— *discors*, 13
— *fulvigula*, 11
— *platyrhynchos*, 10
— *rubripes*, 12
Anhinga, 29
Anhinga anhinga, 29
Aphelocoma californica, 160
Aquila chrysaetos, 96
Archilochus alexandri, 130
— *colubris*, 129
Ardea alba, 32
— *herodias*, 31
Arenaria interpres, 65
— *melanocephala*, 66
Athene cunicularia, 127
Avocet, American, 57
Aythya affinis, 19
— *collaris*, 17
— *marila*, 18

B

Baeolophus bicolor, 175
— *inornatus*, 176
Bittern, American, 30
Blackbird, Brewer's, 247
— Red-winged, 240
— Yellow-headed, 241
Bluebird, Eastern, 186
— Mountain, 188
— Western, 187
Bobolink, 243
Bobwhite, Northern, 113
Bombycilla cedrorum, 198
Bonasa umbellus, 109
Botaurus lentiginosus, 30
Branta canadensis, 5
Bubo scandiacus, 125
— *virginianus*, 124
Bubulcus ibis, 37
Bucephala albeola, 20
Bufflehead, 20
Bunting, Indigo, 223
— Lazuli, 224
— Painted, 225
— Snow, 226
Bushtit, 170
Buteo jamaicensis, 103
— *lineatus*, 101
— *platypterus*, 102
— *swainsoni*, 104
Butorides virescens, 40

C

Cairina moschata, 7
Calidris alba, 67
— *alpina*, 71
— *mauri*, 70
— *minutilla*, 68
— *pusilla*, 69
Callipepla californica, 111
— *gambelii*, 112
Calypte anna, 131
Campylorhynchus brunneicapillus, 180
Cardinal, Northern, 219
Cardinalis cardinalis, 219
Catbird, Gray, 196
Cathartes aura, 92
Catharus fuscescens, 192
— *guttatus*, 189
— *ustulatus*, 191
Certhia americana, 179
Chaetura pelagica, 135
Charadrius alexandrinus, 52
— *melodus*, 53
— *semipalmatus*, 51
— *vociferus*, 54
Chen caerulescens, 4
Chickadee, Black-capped, 172
— Carolina, 171
— Chestnut-backed, 173
— Mountain, 174
Chordeiles minor, 128
Chroicocephalus philadelphia, 75
Circus cyaneus, 100
Coccothraustes vespertinus, 255

Colaptes auratus, 145
Colinus virginianus, 113
Collared-Dove, Eurasian, 115
Columba livia, 116
Columbina passerina, 119
Coot, American, 49
Coragyps atratus, 93
Cormorant, Double-crested, 28
Corvus brachyrhynchos, 162
— *corax*, 164
— *ossifragus*, 163
Cowbird, Brown-headed, 248
Crane, Sandhill, 45
Creeper, Brown, 179
Crow, American, 162
— Fish, 163
Cyanocitta cristata, 159
— *stelleri*, 158
Cygnus olor, 6

D

Dolichonyx oryzivorus, 243
Dove, Common Ground-, 119
— Eurasian Collared-, 115
— Mourning, 118
— White-winged, 117
Dowitcher, Long-billed, 73
— Short-billed, 72
Dryocopus pileatus, 144

Duck, American Black, 12
— Mottled, 11
— Muscovy, 7
— Ring-necked, 17
— Wood, 8
Dumetella carolinensis, 196
Dunlin, 71

E

Eagle, Bald, 97
— Golden, 96
Egret, Cattle, 37
— Great, 32
— Reddish, 36
— Snowy, 33
Egretta caerulea, 34
— *rufescens*, 36
— *thula*, 33
— *tricolor*, 35
Elanoides forficatus, 95
Eremophila alpestris, 169
Eudocimus albus, 42
Euphagus cyanocephalus, 247

F

Falco columbarius, 106
— *peregrinus*, 107
— *sparverius*, 105
Falcon, Peregrine, 107
Finch, House, 253
— Purple, 252
Flicker, Northern, 145
Flycatcher, Great Crested, 149
— Scissor-tailed, 152
Fratercula arctica, 89
Fulica americana, 49

G

Gallinula galeata, 47
Gallinule, Common, 47
— Purple, 48
Gavia immer, 24
Geococcyx californianus, 120
Geothlypis trichas, 202
Gnatcatcher, Blue-gray, 185
Godwit, Marbled, 64
Goldfinch, American, 256
— Lesser, 257
Goose, Canada, 5
— Snow, 4
Grackle, Boat-tailed, 245
— Common, 244
— Great-tailed, 246
Grebe, Pied-billed, 25
Grosbeak, Black-headed, 221
— Blue, 222
— Evening, 255
— Rose-breasted, 220
Ground-Dove, Common, 119
Grouse, Ruffed, 109
Grus canadensis, 45
Gull, Bonaparte's, 75
— California, 78
— Franklin's, 77
— Great Black-backed, 82
— Herring, 80
— Laughing, 76
— Ring-billed, 79
— Western, 81

H

Haematopus palliatus, 55
Haemorhous mexicanus, 253
— *purpureus*, 252
Haliaeetus leucocephalus, 97
Harrier, Northern, 100
Hawk, Broad-winged, 102
— Cooper's, 98
— Red-shouldered, 101
— Red-tailed, 103
— Sharp-shinned, 99
— Swainson's, 104
Heron, Great Blue, 31
— Green, 40
— Little Blue, 34
— Tricolored, 35
Himantopus mexicanus, 56
Hirundo rustica, 168
Hummingbird, Allen's, 134
— Anna's, 131
— Black-chinned, 130
— Broad-tailed, 132
— Ruby-throated, 129
— Rufous, 133
Hydroprogne caspia, 87
Hylocichla mustelina, 190

I

Ibis, Glossy, 41
— White, 42
Icterus bullockii, 251
— *galbula*, 250
— *spurius*, 249

J

Jay, Blue, 159
— Steller's, 158
— Western Scrub-, 160
Junco, Dark-eyed, 239
Junco hyemalis, 239

K

Kestrel, American, 105
Killdeer, 54
Kingbird, Eastern, 151
— Western, 150
Kingfisher, Belted, 137
Kinglet, Golden-crowned, 184
— Ruby-crowned, 183
Kite, Swallow-tailed, 95

L

Lanius ludovicianus, 153
Lark, Horned, 169
Larus argentatus, 80
— *californicus*, 78
— *delawarensis*, 79
— *marinus*, 82
— *occidentalis*, 81
Leucophaeus atricilla, 76
— *pipixcan*, 77
Limnodromus griseus, 72
— *scolopaceus*, 73
Limosa fedoa, 64
Loon, Common, 24
Lophodytes cucullatus, 23

M

Magpie, Black-billed, 161
Mallard, 10
Martin, Purple, 165
Meadowlark, Eastern/Western, 242
Megaceryle alcyon, 137
Megascops asio, 122
— *kennicottii*, 123
Melanerpes carolinus, 139
— *erythrocephalus*, 138
Meleagris gallopavo, 110
Melospiza georgiana, 234
— *melodia*, 235
Melozone crissalis, 229
Merganser, Common, 21
— Hooded, 23
— Red-breasted, 22
Mergus merganser, 21
— *serrator*, 22
Merlin, 106
Mimus polyglottos, 197
Mniotilta varia, 201
Mockingbird, Northern, 197
Molothrus ater, 248
Mycteria americana, 44
Myiarchus crinitus, 149

N

Nighthawk, Common, 128
Night-Heron, Black-crowned, 38
— Yellow-crowned, 39
Numenius phaeopus, 63

Nuthatch, Red-breasted, 178
— White-breasted, 177
Nyctanassa violacea, 39
Nycticorax nycticorax, 38

O

Oriole, Baltimore, 250
— Bullock's, 251
— Orchard, 249
Osprey, 94
Ovenbird, 200
Owl, Barn, 121
— Barred, 126
— Burrowing, 127
— Eastern Screech-, 122
— Great Horned, 124
— Snowy, 125
— Western Screech-, 123
Oystercatcher, American, 55

P

Pandion haliaetus, 94
Parula, Northern, 203
Passerculus sandwichensis, 232
Passer domesticus, 259
Passerella iliaca, 233
Passerina amoena, 224
— *caerulea*, 222
— *ciris*, 225
— *cyanea*, 223
Patagioenas fasciata, 114

Pelecanus erythrorhynchos, 26
— *occidentalis*, 27
Pelican, American White, 26
— Brown, 27
Phalacrocorax auritus, 28
Phasianus colchicus, 108
Pheasant, Ring-necked, 108
Pheucticus ludovicianus, 220
— *melanocephalus*, 221
Phoebe, Black, 146
— Eastern, 147
— Say's, 148
Pica hudsonia, 161
Picoides pubescens, 143
— *villosus*, 142
Pigeon, Band-tailed, 114
— Rock, 116
Pintail, Northern, 15
Pipilo erythrophthalmus, 227
— *maculatus*, 228
Piranga ludoviciana, 218
— *olivacea*, 217
— *rubra*, 216
Platalea ajaja, 43
Plectrophenax nivalis, 226
Plegadis falcinellus, 41
Plover, Black-bellied, 50
— Piping, 53
— Semipalmated, 51
— Snowy, 52
Pluvialis squatarola, 50
Podilymbus podiceps, 25

Poecile atricapillus, 172
— *carolinensis*, 171
— *gambeli*, 174
— *rufescens*, 173
Polioptila caerulea, 185
Porphyrio martinica, 48
Porzana carolina, 46
Progne subis, 165
Psaltriparus minimus, 170
Puffin, Atlantic, 89

Q

Quail, California, 111
— Gambel's, 112
Quiscalus major, 245
— *mexicanus*, 246
— *quiscula*, 244

R

Raven, Common, 164
Recurvirostra americana, 57
Redpoll, Common, 254
Redstart, American, 213
Regulus calendula, 183
— *satrapa*, 184
Roadrunner, Greater, 120
Robin, American, 193
Rynchops niger, 88

S

Sanderling, 67
Sandpiper, Least, 68
— Semipalmated, 69
— Solitary, 59
— Spotted, 58
— Western, 70

Sapsucker, Red-breasted, 140
— Yellow-bellied, 141
Sayornis nigricans, 146
— *phoebe*, 147
— *saya*, 148
Scaup, Greater, 18
— Lesser, 19
Scolopax minor, 74
Screech-Owl, Eastern, 122
— Western, 123
Scrub-Jay, Western, 160
Seiurus aurocapilla, 200
Selasphorus platycerus, 132
— *rufus*, 133
— *sasin*, 134
Setophaga americana, 203
— *caerulescens*, 209
— *coronata*, 206
— *discolor*, 210
— *dominica*, 214
— *magnolia*, 205
— *palmarum*, 212
— *pensylvanica*, 215
— *petechia*, 204
— *pinus*, 211
— *ruticilla*, 213
— *townsendi*, 208
— *virens*, 207
Shoveler, Northern, 14
Shrike, Loggerhead, 153
Sialia currucoides, 188
— *mexicana*, 187
— *sialis*, 186
Siskin, Pine, 258
Sitta canadensis, 178
— *carolinensis*, 177
Skimmer, Black, 88
Sora, 46

Sparrow, American Tree, 231
— Chipping, 230
— Fox, 233
— Golden-crowned, 238
— House, 259
— Savannah, 232
— Song, 235
— Swamp, 234
— White-crowned, 237
— White-throated, 236
Sphyrapicus ruber, 140
— *varius*, 141
Spinus pinus, 258
— *psaltria*, 257
— *tristis*, 256
Spizella arborea, 231
— *passerina*, 230
Spoonbill, Roseate, 43
Starling, European, 199
Sterna forsteri, 83
— *hirundo*, 84
Sternus vulgaris, 199
Stilt, Black-necked, 56
Stork, Wood, 44
Streptopelia decaocto, 115
Strix varia, 126
Sturnella magna, 242
— *neglecta*, 242
Swallow, Barn, 168
— Tree, 167
— Violet-green, 166
Swan, Mute, 6
Swift, Chimney, 135
— White-throated, 136

T
Tachycineta bicolor, 167
— *thalassina*, 166
Tanager, Scarlet, 217
— Summer, 216
— Western, 218
Teal, Blue-winged, 13
— Green-winged, 16
Tern, Caspian, 87
— Common, 84
— Forster's, 83
— Royal, 85
— Sandwich, 86
Thalasseus maximus, 85
— *sandvicensis*, 86
Thrasher, Brown, 194
— Curve-billed, 195
Thrush, Hermit, 189
— Swainson's, 191
— Wood, 190
Thryothorus ludovicianus, 182
Titmouse, Oak, 176
— Tufted, 175
Towhee, California, 229
— Eastern, 227
— Spotted, 228
Toxostoma curvirostre, 195
— *rufum*, 194
Tringa flavipes, 60
— *melanoleuca*, 61
— *semipalmata*, 62
— *solitaria*, 59
Troglodytes aedon, 181
Turdus migratorius, 193
Turkey, Wild, 110
Turnstone, Black, 66
— Ruddy, 65
Tyrannus forficatus, 152
— *tyrannus*, 151
— *verticalis*, 150
Tyto alba, 121

V

Veery, 192
Vireo, Blue-headed, 155
— Red-eyed, 156
— Warbling, 157
— White-eyed, 154
Vireo, gilvus, 157
– *griseus*, 154
– *olivaceus*, 156
– *solitarius*, 155
Vulture, Black, 93
— Turkey, 92

W

Warbler, Black-and-white, 201
— Black-throated Blue, 209
— Black-throated Green, 207
— Chestnut-sided, 215
— Magnolia, 205
— Palm, 212
— Pine, 211
— Prairie, 210
— Townsend's, 208
— Yellow, 204
— Yellow-rumped, 206
— Yellow-throated, 214
Waxwing, Cedar, 198
Whimbrel, 63

Wigeon, American, 9
Willet, 62
Woodcock, American, 74
Woodpecker, Downy, 143
— Hairy, 142
— Pileated, 144
— Red-bellied, 139
— Red-headed, 138
Wren, Cactus, 180
— Carolina, 182
— House, 181

X

Xanthocephalus xanthocephalus, 241

Y

Yellowlegs, Greater, 61
— Lesser, 60
Yellowthroat, Common, 202

Z

Zenaida asiatica, 117
— *macroura*, 118
Zonotrichia albicollis, 236
— *atricapilla*, 238
— *leucophrys*, 237